NINJA INNOVATION

ALSO BY GARY SHAPIRO

The Comeback: How Innovation Will Restore the American Dream

To my sons: Steve, Doug, Mark, and Max. Each wonderful . . . each different. This book is for them and their future children. Ninjas make things happen. And so will they!

HarperCollins books may be purchased for educational, business, or sales promotional use. For information please e-mail the Special Markets Department at SPsales@harpercollins.com.

A hardcover edition of this book was published in 2013 by William Morrow, an imprint of HarperCollins Publishers.

FIRST WILLIAM MORROW PAPERBACK EDITION PUBLISHED 2015.

Designed by Jamie Lynn Kerner

Library of Congress Cataloging-in-Publication Data has been applied for.

ISBN 978-0-06-224233-4

15 16 17 18 19   OV/RRD   10 9 8 7 6 5 4 3 2 1

NINJA INNOVATION

THE TEN KILLER STRATEGIES OF THE WORLD'S MOST SUCCESSFUL BUSINESSES

GARY SHAPIRO

WM
WILLIAM MORROW
An Imprint of HarperCollinsPublishers

CONTENTS

PREFACE

MANY YEARS AGO I ACHIEVED MY BLACK BELT IN TAE KWON DO. THE accomplishment marked the end of a journey that had pushed me to the limits of my mental and physical abilities. The test for every belt on the ladder from yellow to black was an intensely exhausting experience. Your forms had to be perfect, your placements precise, and your discipline unwavering. And when you were finally finished, battered but proud of victory, it started all over again. There was always another kick to learn and another kata to master on the path to a higher belt. The test for the black belt, however, which signifies the repudiation of fear, far surpassed anything I could have imagined.

But my own test was nothing compared to the experience of watching my seven- and eight-year-old sons vie for their black belts. With my stomach in knots, I wanted to rush onto the mat and protect my boys during their sparring matches. But my then-wife and I held steady, because we knew that they were learning an important lesson: accomplishment requires sacrifice and risk.

While all of my family was fortunate enough to receive their black belts, I knew we were in the minority in the class. Many students dropped out after receiving a milestone belt. Others tested repeatedly for black belts but failed over and over. For most, the katas and skills proved too difficult. Sometimes, the board just won't break.

Along the way, I entered a few tournaments, which were scarier and riskier than I had imagined. Practice is one thing; doing it live is something else. But those were the only times I displayed my tae kwon do skills in combat. I have never had to use them in actual self-defense, nor have my sons—to my knowledge at least. This might lead some to wonder: Why would you go through all the personal anguish just for a few colored cotton belts? Good question.

Originally, I studied tae kwon do because I thought it would be a great experience to go through with my sons and then-wife. But as we progressed together, it hit me that this wasn't just about them. I was becoming a more focused, disciplined person. I am not by nature an idle guy, so I didn't need tae kwon do to rejuvenate my passion for work or family. What it did do was shape the way I approached my job; it put everything into a framework that I have since been able to rely on for decades.

If I had to define this framework, it would go something like this: to be successful, one must set goals; to achieve these goals, one must form a strategy; and to fully execute that strategy one must never let failures get in the way. Indeed, one must use those failures to get better. I learned this simple formula through my study of tae kwon do, and I've continually applied it to my career long after hanging up my *gi*.

But there's another element to this framework that defies a simple definition. It is innovation. To be successful, one has to move beyond what has come before. One must take the lessons learned through study, experience, and failure, and apply them in ways that change the game, so to speak. You cannot be successful in tae kwon do if all you do is repeat what you are taught. A beginner white belt learns this early, when he or she must spar for the first time. Your competition won't play fair; it won't do what you expect; and it will always, always fight back. When you're on the mat, you must assume your competition knows what you know; that it too has a strategy for victory. The only way you can defeat your opponent is if you do something he does not expect; you must innovate or die.

This happens to also be the fundamental lesson I have learned from thirty years working and leading the most dynamic industry on the planet: the consumer electronics industry. You could measure our success through dollars and cents—and indeed, these are impressive in their own right. But that's not how I measure our achievements. For me, as for all of our members, success is defined by reaching a previously insurmountable height. What's been remarkable is that CE companies have done this again and again. The world that we live in today is unrecognizable to that of just twenty years ago. From smart phones and e-readers to HDTV and wireless broadband, from digital cameras and MP3 players to tablet computers and motion-detection gaming—nothing is ever static; no product is ever immune to competition; and we can never tell where we'll be in five years. The simple reason: innovation.

Which is not to discount other highly successful companies in other industries, some of which I will discuss in the following pages. But my experience has been with consumer electronics and I have been fortunate to witness this industry develop into the colossus that it is today. And as I've watched my industry rise, I've seen some of the best business models ever devised. I've also witnessed a fair share of pretty bad ones. I've observed companies shine as bright as a shooting star, only to flame out just as quickly. I've seen other companies rise and fall and rise again, like the proverbial phoenix. And then there are those companies whose only trajectory is ever upward, whose employees and products are part of the American tale.

Not so long ago it struck me that these stories were worth telling. Particularly during a time of economic uncertainty, where millions of Americans are out of work and pessimistic about our future, I realized there was a need to remember what made us into the greatest nation on earth. In 2011, I published *The Comeback: How Innovation Will Restore the American Dream*. In it, I laid out a strategy for the country that would not only put innovation at the heart of U.S. economic policy but also reinvigorate America's great innovative spirit.

The Comeback outlines the fundamentals that can lead to economic success on a national scale. In other words, at a very high, stratospheric level. What about the individuals and enterprises *on the ground* whose collective actions have built America and will fuel our national resurgence? Surely, I thought, there are some lessons we can learn (or remember) from those who changed our lives in both good economic times and bad. After all, I must have some notion, considering that for thirty years I have worked in an industry whose successes and failures are the stuff of legend.

It was in analyzing that question that this present book came to be. The challenge was to find the common traits that connected these successes with one another. A book that simply documents successful companies without asking what made them so successful wasn't particularly interesting to me. Then I recalled my own framework for success and wondered if it would apply to the stories I wanted to tell. I learned my framework through the study of tae kwon do—which, stripped to its core, is the study of how to be a warrior. But not just any warrior, rather a highly disciplined, strategic, and passionate warrior. In other words, a ninja.

In the following pages, I hope to explain better what I mean by the term *ninja innovation*. What do these ancient warriors of feudal Japan have to tell us about success in business? What can we learn from them that might help rejuvenate our passion for innovation? It is this connection—between the ancient ninja warrior and today's innovation warrior—that is the purpose of this book. It is the story of the entrepreneur who toils away at an idea; it is the story of a CEO who saves a company on the brink of collapse; it is the story of the enterprise that has everything going for it, yet fails anyway; it is the story of a corporation that time and again reinvents itself to stay dominant. What ties these tales together? What defines them? What separates them? What can we learn from the successes as well as the failures?

Ninja Innovation will provide the answers.

INTRODUCTION

THE WAY OF THE NINJA

Nothing Has to Happen

IN JUNE 1967, THE FIRST TRADE SHOW DEVOTED TO CONSUMER electronics was held in New York City. At the time, consumer electronics (CE) was an $8 billion industry ($55 billion in today's dollars) that centered on the TV and hi-fi markets. But the industry was also experiencing something of a downturn, as radio sales were declining sharply, and several of the industry's major players had made highly publicized layoffs.

It was in this uncertain environment—a growing but not yet secure CE industry—that Jack Wayman, an Electronic Industries Association (EIA) executive, determined that a specialized CE trade show was necessary. And so in the early summer at the Americana and Hilton Hotels in midtown Manhattan, fifteen thousand manufacturers, distributors, and retailers walked among a hundred exhibits occupying a hundred thousand square feet and featuring more than a thousand new products ranging from "$8 radios the size of a pack of cigarettes to $15,000 room-length stereophonic units," according to coverage of the show in the *New York Times*.

On display were a wide variety of TVs, transistor and tabletop radios, record players, and console furniture units that combined all these advanced audiovisual (AV) components. The latest tech rages were stereo audiotape player/recorders that used "cartridges" or "Playtape"—the then-new compact cassette format—and a growing number of devices that relied on solid-state components, including integrated circuitry rather than vacuum tubes. According to the *Times*, there was "a television set, radio, or tape recorder to suit any taste or price."

Fast-forward forty-five years. That specialized trade show for the CE industry had grown into the International CES (Consumer Electronics Show), the largest trade show in the Americas. Here's a sample of one of dozens of stories the *Times* filed from Las Vegas at the 2012 International CES: "There were a million gorgeous new Android phones and Windows phones, many of them 4G (meaning faster Internet in big cities)," one report read, with slight exaggeration.[1] "Microsoft revealed that its popular Kinect, which plugs into an Xbox and lets you play games just by moving your arms and legs in front of the TV, will now be available for Windows computers."

Android, Windows, Kinect, Xbox, 4G? To a time traveler from 1967 these words would be meaningless. But we know immediately what they are: a cell phone, a computer operating system, video game consoles, and the term for ultra-broadband Internet access. In 1967, CES attendees marveled at $8 tape recorders. Today, we marvel at pocket-sized devices that carry more computing power than the Apollo rockets that took Americans to the moon.

Indeed, the 2012 International CES was the largest to date—with more than 156,000 attendees and 3,100 exhibitors. Meanwhile, in 2012 U.S. factory sales of consumer electronics exceeded $200 billion annually, and worldwide sales topped $1 trillion for the first time. That's a 1,700 percent increase from 1967.

In 1967, the big technological advance was Ray Dolby's noise-reduction system. In 2012, the killer innovation is . . . well, where to begin? Technological breakthroughs happen so frequently that to argue one is bigger than the other is, practically speaking, not a useful exercise. Can we say that Apple's iPad is more advanced than Amazon's Kindle Fire or Samsung's Galaxy? Some say yes, some say no. What's astonishing is the remarkable pace of innovation set by the most successful consumer-electronics companies.

As it was after World War II, after the recession in the late 1970s, and after the recession in the late 1980s, once again it looks as if CE devices will be a prime economic generator to help the economy emerge from the current downturn. Despite the recession, global spending on consumer technology devices topped a record $993 billion in 2011. And for the first time in history, the world's most valuable company—Apple—sells not cars or oil, but consumer electronics.

In the previous CE-fueled recoveries, consumer gadget spending was spurred by "must-have" technology that was barely a fantasy—if even that—only a few years earlier. After World War II, for instance, everyone wanted that newfangled thing called a television, along with new hi-fi audio gear and the transistor radio. After the oil-embargo-induced recession in the mid-1970s, enthusiastic uptake of new videocassette recorders (VCRs), personal computers (PCs), and compact discs (CDs) boosted the economy. Following the stock market crash in 1987, cordless phones, cell phones, personal digital assistants (PDAs), the World Wide Web, digital cameras, satellite TV, Global Positioning System (GPS) devices, and high-definition television (HDTV) spurred the 1990s tech boom.

This time around, we are rushing online and into stores to snap up touch-screen devices, gesture-based game consoles, Internet-connected "smart" HDTVs and Blu-ray players, 4G smart phones more powerful than desktop PCs were a decade ago, e-book readers, media streaming set-top boxes (STBs), cloud-based subscription services, and Ultrabook laptop PCs measuring less than an inch thick and weighing fewer than three pounds.

While we can't yet buy the personal jetpacks or *Star Trek*–style transporter chambers that many futurists thought we'd all have here in the twenty-first century, the next few years will bring even more new technology into our homes and businesses—organic light-emitting diode (OLED) HDTVs as thin as a pencil and as light as a Thanksgiving turkey; 802.11ac Wi-Fi with speeds measured not in megabits per second but gigabits; Thunderbolt device interconnections to complement and finally replace USB; "super" Wi-Fi hotspots measured not in feet but in miles; ultrawide-screen 21 x 9 HDTVs; body sensors to keep a constant eye on our health; and maybe even driverless cars.

While a CE-fueled recovery has historical precedent, at no time in gadget history has the influx of new foundational technologies been so numerous and impactful. In other words, for all its resemblance to 1967, we live in a world that is profoundly different. This isn't solely because of the revolutionary products that have emerged from the industry I represent (although I am biased); nor would I attempt to argue that we live in a *better* world. There are too many other factors that go into what one would deem *better*.

Rather, these products have become *indispensable* in our lives. An American living in the second decade of the twenty-first century cannot lead a fully productive life without a cell phone. A twenty-first-century company cannot survive without giving every one of its employees access to the Internet. And, to be a bit over-zealous, the global economy would implode if suddenly the World Wide Web disappeared. Nearly every modern system, private and government, requires the Internet—with the possible exception of the U.S. Postal Service, which would probably see the Web's disappearance as the dawning of a golden age.

The reason I summarize this progression is not because this book is a history of the consumer electronics industry; it isn't. Rather, I raise this point because we have a tendency to look at an indispensable tool—be it electricity, an iPhone, or even a car—and think it exists because it must exist. In other words, we take these tools for granted after the honeymoon with them is over. And we assume that they came into our lives for the reason the sun rises in the east: It is the way it was supposed to be.

What we forget is that it wouldn't have been this way unless an individual person, company, organization, or country made it happen. Technological progress is not on an inevitable trajectory pointing ever upward. Nothing *must* happen. Take the most revolutionary technology to hit humanity since the harnessing of electricity: the Internet and the World Wide Web.

The Internet was developed in 1969 by ARPANET (the Advanced Research Projects Agency Network), which was founded by the U.S. Department of Defense in 1958 and charged with advanced technology research in the wake of the Russians launching Sputnik. The idea was to provide a decentralized communications network that would not be disrupted by potential global war. Any individual node of the network could be knocked out, but these destroyed nodes could be bypassed and the network would survive. Academia became the largest user of the Net, with researchers linking up their computers to share data.

Right here, we can begin playing the game of what-if. What if the Russians had never launched Sputnik, sending the United States into a technological panic? What if someone had made a different strategic decision at the Defense Department? Would the Internet have been created anyway? We like to think so. But there's more to the story.

In 1989, Tim Berners-Lee and a group of fellow researchers at CERN, an international scientific organization based in Geneva, Switzerland, created a computer code called hypertext transfer protocol (HTTP), a text format code called hypertext markup language (HTML), and a universal resource identifier (later universal resource locator, or URL) for identifying document locations, which formed the basis for the World Wide Web. In 2010 and 2011 respectively, the two men who developed transmission control protocol (TCP) and Internet protocol (IP), Dr. Vint Cerf and Dr. Robert Kahn, were inducted into the Consumer Electronics Hall of Fame and named Consumer Electronics Association Digital Patriots in honor of their contributions to the development of the Internet.

If the U.S. government created the framework, then CERN built the rooms and elevators. A few years later, companies like CompuServe, AOL, and Prodigy were the first tenants. So roughly thirty-five years passed from the "invention" of the Internet to when the first American heard the famous words "You've got mail." To say that all that—all the decisions, people, circumstances, and sheer luck—would have happened no matter what is taking too much for granted.

The Internet is a success story of monumental proportions, if only because it all could have been so very different. And that's what this book is about: a uniquely powerful and democratic way of ensuring success. In my thirty years working within the consumer electronics industry, I have seen more than my fair share of successes—and I've seen many failures. The reason I provided a brief summary of the industry I know best is because it is not the story of the inevitable momentum of history. Rather, it is the story of a thousand different successes, some isolated in their occurrence, and some intimately connected to others. Today, we look at the dominance consumer electronics hold over our lives but we rarely wonder if it could have been any other way.

The answer: Yes, it could have. I know because I saw what it took for these revolutionary products to change our lives. It was not easy. The conceit of this book is that I have in my thirty years' experience gleaned some insight into what it took for these individuals, companies, and societies to achieve success. Were I the head of the association representing the U.S. steel industry, perhaps my insight wouldn't be worth much. As it is, I was fortunate to be present at the creation, so to speak, of the most impactful industry driving America's economic progress. I take no credit for these successes; I am merely relating what I saw and what I learned. I think it's a story worth telling. My hope is that you agree.

Discipline and Goals

As I mentioned in the preface, my inspiration for this book began when I started thinking about the essential traits of a successful person, company, and organization. What ties them all together? Surely, there must be some shared qualities that lead one person to succeed while his peer fails. Then I recalled my study of tae kwon do.

Studying martial arts teaches discipline. Simply the act of regularly attending class reinforces the value of a disciplined approach to personal growth and development. Additionally, the classes themselves include rituals that help develop the focus, respect, and inner resolve necessary for personal success.

Upon entering the school or greeting an instructor, students are taught to put their hands at their sides and bow respectfully. After a few times this bow becomes a habit, but the ritual underscores the importance of showing respect to the teacher and the school.

Each class begins with the whole class reciting a pledge together. While the pledges certainly vary among schools and types of tae kwon do, most share common themes of respect, discipline, and basic ethical principles. My school's tae kwon do pledge was as follows:

> WE COMMIT OURSELVES TO MENTAL AND PHYSICAL DISCIPLINE
> TO BE FRIENDS WITH ONE ANOTHER AND TO DEVELOP STRENGTH IN OUR GROUP
> WE SHALL NEVER FIGHT TO ACHIEVE SELFISH GOALS
> BUT TO DEVELOP WISDOM AND CHARACTER IS OUR ULTIMATE COMMITMENT.

Saying the pledge each day before class reinforced our shared goals and helped build our personal discipline. More, it was part of the ritualistic comfort that becomes ingrained in any communal activity. As when schoolchildren say the Pledge of Allegiance each morning, repetition is good for daily reflection and character building.

The recitation of the pledge was followed by the class schedule of intense and exhausting warm-ups, including stretching and push-ups. We would then review katas—specific stylized routines of kicking, punching, and blocking. The katas become more advanced the farther a student progresses, but their purpose never changes: Routine is good for discipline.

We would also practice kicks, receive weapons training, and conclude with sparring. For the sparring we would wear protective clothing, gloves, and masks, and focus on hitting with control. As a parent, I watched with concern as my young sons sparred, but the pads and helmets were enough to absorb the blows, which gives the combatant the opportunity to focus.

Reaching the next belt level—a constant goal in tae kwon do—naturally reinforces all these habits as you slowly but surely learn the art. It's not just physical progression either, although it is a great way to stay in shape; also it's about building a mental toughness that you can use in other life pursuits. It is a discipline that requires discipline!

The entire experience helped my sons and me develop discipline, respect, and self-confidence. While I don't think I can take on Jet Li just yet, I can say with assurance that I am a more focused person because of my martial arts training. But perhaps most important, studying martial arts is about setting and achieving goals—and that struck me as a perfect metaphor for what I've witnessed in my career with the most successful people: They are goal setters, they are achievers, they strive for the highest rung, and even if they don't reach it at first, they try again.

But *martial arts innovation* doesn't have the best ring to it. *Ninja innovation?* Much better.

NINJA INNOVATION

The Way of the Ninja

IT WASN'T JUST THE NAME, EITHER. AS I THOUGHT ABOUT IT, I REAL-ized that while martial arts is an art, being a ninja is a profession: A ninja is a particular brand of person who studied the martial arts, but not simply for the sake of doing so. They studied so that they could be better ninjas. This seemed to me a key differentia-tor, and so I began to study the way of the ninja. Who were they? What were their roles? Why have they persisted in popular legend hundreds of years after they disappeared from history?

I think a key to understanding ninjas is to compare them to their feudal Japanese counterpart: the samurai. Now, the samurai were akin to European knights. One was born into the samurai class, just as knights were members of feudal Europe's landed ar-istocracy. The samurai were noble warriors who followed a code of ethics and conduct—known as Bushido. Indeed, they took their role as masters quite seriously, and dishonor was a great crime among the samurai class.

Shiba Yoshimasa, a fourteenth-century samurai, stated that the ultimate glory for a samurai was death in battle: "It is a matter of regret to let the moment when one should die pass by. . . . First, a man whose profession is the use of arms should think and then act upon not only his own fame, but also that of his descendants. He should not scandalize his name forever by holding his one and only life too dear. . . . One's main purpose in throwing away his life is to do so either for the sake of the Emperor or in some great undertaking of a military general. It is that exactly that will be the great fame of one's descendants."[2]

Very brave, but anyone who follows that type of strict ethical code isn't exactly concerned about success. In other words, the

10

samurai followed their code to a fault. While they were very interested in self-preservation through the expansion or defense of their feudal power, they weren't exactly goal-oriented.

Let's contrast this with the ninja. Little is known about their exact origins, but they began to appear around the fourteenth and fifteenth centuries as spies for the samurai. They were recruited from the lower classes—another telling detail—and valued for their unique skill set and training. As opposed to the samurai, who were born samurai whether they were any good with a sword or not, the ninja lived in a meritocracy: He could only advance as far as his talents could take him. In time ninjas would expand their skills into what we know today as unconventional warfare, excelling at the art of espionage, sabotage, infiltration, and assassination. After existing mostly in the shadows for many generations, ninjas by about the sixteenth century were a well-known class of warriors whom the varying clans of feudal Japan hired to do their dirty work.

Whereas the samurai fought honorably in open combat, a ninja would just as soon slit your throat while you were sleeping. Not very honorable, but it got the job done. While the ninja followed a code of honor and discipline, he didn't much care if he died a "good death" in the course of his mission. He only cared if the mission was completed. But it's a mistake to think of ninjas as feudal Japanese Terminators—mindless killers. As one historian wrote of a group of ninjas: "They travelled in disguise to other territories to judge the situation of the enemy, they would inveigle their way into the midst of the enemy to discover gaps, and enter enemy castles to set them on fire, and carried out assassinations, arriving in secret."[3]

And so the ninjas had to be smart as well as adept professionals. They had to survey the defenses of their enemy and discover how to beat them. And they couldn't just bring the mightiest force to the battlefield to do so. That wasn't their job. They had to look at a fortified position, like a castle, and investigate its weaknesses; avoid detection; change course if surprises arose; and, finally, complete the mission they were assigned. In short, ninjas had to be innovative in besting their competition.

What better description is there for a successful person, enterprise, or organization in today's world? And so: *Ninja Innovation*.

As for the innovation part, readers of *The Comeback* might recall what I mean by the word, which encompasses far more than simply "an invention." When I say *innovation*, I'm talking about progress. I'm talking about growth. I'm talking about the essential element every successful organization needs not just to survive, but to thrive. If you're doing it like everyone else, you're not innovating. Indeed, sometimes innovation comes in the form of a flashy new product, but not always. Sometimes it comes in the form of a new way of doing business. I put innovation at the pinnacle of economic performance because without it, we are idle. With it, we achieve great success.

But for the business readers in the audience, we can try to be even more precise. At the most basic level, there are three principal types of business innovation[4]:

- Evolutionary: an improvement in an established market that competitors and customers generally *expect* to happen. For example, I would say that faster computer chips are evolutionary.
- Revolutionary: an improvement in an established market that competitors and customers generally *do not* expect to happen. For example, the introduction of smart phones was a revolution in the mobile phone market.

- Disruptive: an improvement that is generally unexpected by customers and competitors, serves a new set of customer values, and ultimately creates a new market that competitors scramble to understand and adapt to. If any innovation warrants stealth development, it's probably this type. For example, the advent of mobile telephones was disruptive to the traditional landline telephone market.

Ninja innovation is my catch-all phrase for what it takes to succeed. You have to display the qualities of the ancient Japanese ninja, whose only purpose was to complete the job. He wasn't bound by precedent; he had to invent new ways. He didn't have the luxury of numbers; he had to make do with a small group of professionals. He wasn't asked to do the ordinary; he had to perform extraordinary tasks.

Of course a major difference between the ancient ninja and today's ninja innovator is that for the former, failure was fatal. That isn't true today. One can fail, dozens of times in fact, before finding remarkable success. Indeed, America is quite unique in the world because we actually reward failure: We want you to go down trying. Some of our most spectacular successes were also spectacular failures: In war, George Washington failed countless times against the British before achieving miraculous victory; in politics, Abraham Lincoln lost his Senate race to Stephen Douglas—while giving us some of the most memorable debates in history—before saving the Union; in industry, Henry Ford's first company, the Detroit Automobile Company, dissolved three years into its existence, but Ford Motor Company launched two years later; and, perhaps best known to us, Steve Jobs failed probably more often than he succeeded. And, yes, I have failed more times than I care to recall.

Today's Ninjas

FOR THE PURPOSES OF THIS BOOK, THE ANCIENT-NINJA METAPHOR will only carry us so far. Which is another way to say that what follows is my interpretation of what today's ninjas exhibit: how they see their task, how they tackle the competition, how they overcome the odds, how they plan, how they fight, and how they ultimately succeed. After all, this is not meant to be a history course on ninjas, but an examination of success in today's world.

The chapters that follow will attempt to break down the ten characteristics of ninja innovation. In brief, they include:

Chapter 1, "Your Goal Is Victory": The goal of a ninja is to defeat the enemy and complete the job. Likewise, the goal of an enterprise is to be better than the competition. This chapter will examine cases and examples where enterprises succeeded because they were driven to win, and where the ultimate goal of the business strategy was clearly defined as victory.

Chapter 2, "Your Strike Force": Ninjas often operated as a team. More important, they were a team of professionals, not amateurs. One of the first steps toward success in any enterprise is to build the right team.

Chapter 3, "In War, Risk Is Unavoidable": Ninjas, and successful leaders, approach their assignments as a way of life, not merely as a "day job." If you don't take risks, you won't be successful.

Chapter 4, "Prepare for Battle": A ninja's behavior is grounded in a specific mental attitude. We can call this discipline. To succeed, you must mentally steel yourself for the trials ahead. You will fail, often spectacularly. But never take your eyes off your goal.

Chapter 5, "The Art of War": Successful strategy is an art, not a science. Often you will not know all you should know to achieve success. That's okay. A successful strategy is a living strategy; it must be executed in a way that allows for a change in tactics. Your competition is fierce and smart, and they won't willingly let you defeat them. Expect surprises and adjust accordingly.

Chapter 6, "The Ninja Code": No matter the goal, all actions are "informed" by a martial code of conduct, a.k.a. business ethics. Ninjas create chaos because they don't follow the normal rules; that's how they succeed. But even ninjas follow a code of ethics.

Chapter 7, "Ninjas Break the Rules": Unlike their feudal counterpart, the samurai, ninjas were not an aristocratic class. They succeeded because they were the best. Likewise, an organization won't succeed if its hiring principles are hereditary, hierarchical, and closed to mavericks. The last ninja standing is the enterprise or individual that not only employed the best people but also pursued the most innovative approach to success.

Chapter 8, "Innovate or Die": The ancient ninja always confronted obstacles that forced him to alter his approach. It's the same with today's enterprises. Life isn't predictable. But too often dying organizations turn to third parties (usually government) to save them without having the courage to change their mode of business. Be creative; be daring; be willing to take a different course. Otherwise, your failure *will* be fatal.

Chapter 9, "An Army of Ninjas": Today's ninjas are part of something larger than themselves. As they build, they also defend. Technology allows for all of us to participate in innovation.

Chapter 10, "The Shadow Warrior": The ninja's best skill was stealth. He was able to deceive his enemies through invisibility and disguise. Great ninja companies do this as well, but this is the one ninja trick that isn't part of the innovator code.

As I mentioned previously, *The Comeback* analyzed the macro factors that foster innovation and economic growth. *Ninja Innovation* focuses on the micro factors that lead to individual and organizational success. But both remain forever connected. With a few notable exceptions, the successes I recount in the following pages would not have been possible if not for the freedoms of capitalism and the pro-innovation, pro-growth policies of government. We remain in need of these today.

CHAPTER ONE

YOUR GOAL IS VICTORY

"I'm going to destroy Android, because it's a stolen product." So said Steve Jobs to his biographer Walter Isaacson shortly before his death. Imagine that. Jobs knew he had not long to live and yet there he was, still trying to win. That's a ninja for you.

It's no surprise to say that Jobs possessed many qualities of the ninja innovator. Jobs was smart, passionate, and relentless. A ninja must be all these things, but he must be something more: He must be driven by an unyielding desire to win. A ninja's job is not complete until he finishes the mission. Just infiltrating the castle is meaningless unless he can get back out. For Jobs and other ninja innovators, just inventing a good product is not enough. What is required is different for every person, but one thing is the same: To complete the mission, your goal must be victory.

What also made Jobs a ninja is that he never retired while he could still produce. And so there he was, in the twilight of his life, vowing to "destroy" the competition. It's an appropriate word choice for a famously demanding person. His example inspired countless people to pursue victory in their own lives, if not necessarily in the same way. Not only that, but Jobs's products, and their successors, will be with humanity for centuries—much as another ninja innovator's products are with us today.

Useful, Not Nifty

ON NOVEMBER 14, 2007, CON EDISON ENDED 125 YEARS OF DIRECT-current electricity service that had begun when Thomas Edison opened his Pearl Street power station in Manhattan on September 4, 1882. While the station itself clearly had changed over the years, that's as if in 2003 Ford Motor Company had finally stopped running the assembly line Henry Ford had first fired up in 1903. It's a remarkable testament to the brilliance of Edison.

Today, Edison is known as the innovator of the phonograph and incandescent light. But those weren't his greatest contributions. It was his development of a central power station that has resulted in his most enduring impact on the future of consumer electronics.

The first thing you do when you buy a stereo, TV, or any electronic device is plug it into a wall socket. This simple act is possible because Edison spent almost his entire fortune and risked his substantial reputation to build the Pearl Street power station in lower Manhattan. When Edison gave the order to flip the switch on September 4, 1882, stores all along Fulton and Nassau Streets, the editorial offices of the *New York Times*, and the brokerage house of Drexel Morgan were lit up with Edison's incandescent lightbulbs. Imagine that moment, when Edison knew he had achieved victory. He had forever changed Manhattan—and the world.

The reason the Pearl Street station was so important—and why Edison would have failed had he not created it—is that the electric lightbulb couldn't replace the gas lamp as the primary source of lighting until an entire electrical system was created to sustain it. Otherwise, Edison would have been just a guy who had created a cool, but useless, gadget. In other words, inventing the lightbulb was not the end for Edison; it was only the beginning. The end would come when homes were lit by the lightbulb—and the effort to make that happen proved even more demanding and expensive than the invention of the lightbulb ever was.

The story of Edison and his Pearl Street power station defines for me what I mean by "Your goal is victory." No one questioned that Edison had brilliance, but there are thousands of others out there inventing nifty things that we will never ever see. And Edison wasn't interested in just inventing nifty things. He was driven to win—and for him that meant creating something of value to everyday people, something that, yes, would revolutionize the world, but also something that would require no more from the user than simply flipping a switch. But to go from the lightbulb to turning on the switch required an expenditure of time and resources that would have surpassed the ability of most other inventors.

The same could be said of Jobs. He wasn't interested in creating a cool-looking device that was only used by the wealthy or technology geeks. He wanted to create devices that were useful to everyone. What this requires is a resolve and discipline that so often separate victory from defeat. Having brilliance just isn't enough. History is strewn with the leftovers of otherwise brilliant people who, for whatever reason, just couldn't succeed. Maybe they were just unlucky. But what is certain is that to win, you must have a clear idea of what it fully takes to win and where that requires you to go. For Jobs and Edison, their vision matched their determination to win.

Adapt, Adjust, Dominate

THE DRIVE FOR VICTORY ISN'T JUST EMBODIED IN THOSE RARE INDIviduals who possess a seemingly superhuman intellect, vision, and determination to succeed. It's also something that must propel an enterprise, whether it's an organization like the Consumer Electronics Association (CEA), an athletic team like the New England Patriots, or a company like IBM.

Let's look more closely at IBM. Around the same time Jobs and Steve Wozniak were starting Apple, Paul Allen and Bill Gates set up shop in Albuquerque, New Mexico, to write computer programs, including a new computer operating system, PC-DOS, for the IBM Personal Computer. Gates and Allen and their new company, Microsoft, then adapted PC-DOS into MS-DOS for use in a PC that IBM was trying to build in order to capture some of the non-mainframe computer market opened by Apple.

By 1981, several companies, such as Atari, Tandy, Sinclair, Commodore, and others, had introduced small PCs that were of interest only to relatively niche market segments. This changed when several entrepreneurs introduced more functional computers that ran software such as MicroPro's WordStar (1978) and the SuperCalc spreadsheet program (1980). Adam Osborne's Osborne 1, a twenty-four-pound device with a five-inch screen, was the first "portable" PC when it hit stores in June 1981. Only 8,000 were sold that year, but sales jumped to 110,000 the following year. At one point, Osborne reported an order backlog of 25 months, but the company declared bankruptcy in September 1983. The Osborne 1 was followed by the Kaypro II transportable in late 1982, a similar bulky machine that experienced similar sales success.

But the real coming of age for the PC came between the introductions of the Osborne 1 and the Kaypro II: the unveiling of the IBM PC model 5150 in August 1981.

The IBM PC used Intel's new 8088 chip, a wealth of off-the-shelf computing technology, and, most importantly, the MS-DOS operating system provided by Microsoft. IBM, the world's largest computer maker, quickly proved that there was a substantial market—at least among mainstream businesses—for the PC. Unlike Apple, IBM aimed its PC at the market segment it was more familiar with, and indeed already dominated: the business office. It would prove to be a wise decision.

A host of companies such as DEC, NEC, Xerox, Epson, AT&T, and HP—none of which previously believed that a mainstream, non-geek-oriented computer market existed—quickly jumped on the PC bandwagon. These machines came to be known as "IBM clones" since they essentially copied the technology included in the IBM PC, most of them running MS-DOS. In late 1983, IBM itself tried to expand beyond the market it had created by releasing its IBM PCjr, designed for the home market. But a lack of features and capabilities for average homeowners (and its much-derided "Chiclet" keyboard) doomed the product. Yet IBM's failure was short-lived because it became better at understanding the needs of new and broader market segments.

IBM followed up the PCjr in 1984 with the Kaypro-like IBM "transportable." A start-up company called Compaq made the first DOS-compatible transportable in 1986. The laptop age had begun, and today it not only survives, but it also arguably birthed myriad other portable personal computing devices. But it wasn't a straight line of technology or market development, much as the wars that ninjas fought often ebbed and flowed over decades.

Monochrome computer screens slowly gave way to color screens in the mid-1980s, thanks in part to the introduction of the 256-color VGA monitor included with the IBM PS/2 line in 1987. When videodiscs emerged in the early 1990s, with several formats available, there was serious threat of a format war. To avoid an industry-killing battle, the two competing companies, Sony and Toshiba, enlisted IBM to lead a multicompany committee called the Computer Industry Technical Working Group (TWG) to mediate a solution. Along with IBM, Apple, Microsoft, and HP, other members included Compaq, Sun, Kodak, and Intel. IBM executive Alan Bell chaired the group.

In 1996, IBM designed a chess-playing system known as Deep Blue. Smarter than any previous computer player—the computer player on your PC is a nitwit comparatively—Deep Blue (and IBM) challenged the world champion Garry Kasparov to a match. Kasparov went on to win. But a year later Deep Blue was back and IBM challenged Kasparov again. IBM wanted to win, which it did in a six-game match with two wins to one and three draws.

In 2005, IBM sold its PC division to Lenovo. Three years later, the company's Blue Gene supercomputing program was awarded the National Medal of Technology and Innovation by President Barack Obama. Then, in 2011, an introduction known to millions of people around the globe would forever be associated with IBM: "This is *Jeopardy!*"

In February 2011, during a three-episode tournament, the two most successful champions in *Jeopardy!* history, Brad Rutter and Ken Jennings, competed against a player named Watson. After the first round of the two-game tourney, Jennings had $4,800, Rutter had $10,400, and Watson had won $35,734. After the second game, where the players' scores carried over, Rutter came in third, with $21,600; Jennings came in second, with $24,000; and Watson won the tournament with $77,147.

Of course, as everyone knows, Watson was not a human player but an artificial-intelligence computer system able to answer questions posed in natural—as opposed to computer—language. And compared to Deep Blue's slight victory over Kasparov, Watson's was crushing. The AI did have some unfair advantages over its human competition, such as response time, but its ability to answer questions stunned the world.

Jennings, using the humor that only a human (for now) can, summarized everyone's reaction: "I for one welcome our new computer overlords." IBM had triumphed once again.

Not bad for a company founded in 1911 under its original ninja name, the Computing-Tabulating-Recording Company, or C-T-R. In its 102 years—and I only summarized the last 35—IBM has successfully reinvented itself over and over to the point where it is the second-largest U.S. firm behind Wal-Mart in terms of number of employees.[1] In 2012, *Fortune* magazine ranked IBM the fourth-largest company in terms of market capitalization, the ninth-most profitable, and the nineteenth-largest firm in terms of revenue.[2] That success is largely due to the ninja strategy that IBM has implemented through savvy product positioning and innovation. IBM has created software and programs to run cities and businesses by absorbing data and information and leveraging it to increase profitability, inventory management, planning, and efficiency.

IBM could have appeared in almost any chapter in this book,

but I included it here because IBM is a company whose strategic goal has always been victory. To achieve victory, despite a *Titanic*-like size that might make it difficult to alter course nimbly, IBM has successfully transformed itself from a mainframe company to a PC company to a consulting company to a company that beats chess and *Jeopardy!* champions just because it can. It takes a ninjalike discipline, long-term vision, and technical and marketing skills to be a consistent winner. But more than that, it takes a desire to win.

IBM has rarely been defeated by its competition, but when it does lose—as with its PC division—it has the discipline to cut its losses and move to a new battlefield. Although the world celebrates both Microsoft and Apple in the early PC age, in many ways it was IBM that made it all come together, as the above history makes abundantly clear. The reason I recounted the early years of the personal computer is to highlight that while sometimes market victory is explosively quick, it can also be achieved with more modest, tightly calibrated successes. Successful ninjas, going to battle with limited resources, had to have a cold-eyed realism about what it would take to ultimately succeed, and short-term defeats were lessons from which to learn.

For instance, while Apple initially trounced the competition with the Apple II and then the Macintosh, it was IBM that stole the lead in the late 1980s and throughout the 1990s—of course with a little help from Microsoft. Part of the reason for Apple's early success—and it's certainly not the only one—is that it chose to target families with a user-friendly product the entire family could enjoy. Apple made its Macs easier to use than any IBM-based PC, and it began to dominate the youth market by capturing customers first in their schools.

But then IBM, with an improving Windows operating system, altered its strategy. If a ninja knows that the wall blocking him from completing his mission is impassable, with a wide, deep moat; sheer sides; and hundreds of sharp-eyed archers on the barricades, he has two choices: either try to scale the wall and hope for the best, or find a way around it. Given a competitor's product that was arguably superior in the home market segment, IBM faced a similar choice: attempt to compete in the home market segment with an inferior product, or instead avoid direct engagement by attacking a different, but related, segment. Critically, what defined victory would remain unchanged: controlling the PC market. It would be the tactics that made the difference.

IBM's PCs were very successful in the huge business market and therefore were intimately known to one of, if not the, key decision makers for choosing a PC for the home: the full-time breadwinner, whether father or mother. IBM knew this market; this market trusted IBM. Like any good ninja, IBM simply went around the "wall."

IBM's strategy helped it remain the dominant competitor in the PC market, but it didn't "destroy" Apple. After the second coming of Steve Jobs, Apple is bouncing back, but its share hasn't risen above 10 percent or so for nearly twenty years. IBM's later decision to spin off its personal computer division to Lenovo proved prescient: Both Hewlett-Packard and Dell faced layoffs in 2012 because of declining PC sales.[3] The computing market is moving toward mobile technology.

IBM's experience with the PC offers several lessons regarding ninja strategies and tactics. The first is that your goal (victory!) almost never depends only on you, but instead often requires the aid of allies. These allies include early-adopter customers, technology partners, distribution partners, and so on. Your relationship with them survives only at the point of interaction for your shared interests. The reality of this "what's in it for me?" relationship means that an alliance is rarely permanent. Therefore you should plan in advance for the best way to dissolve it as well as to possibly reconstitute it at some future point.

Another major lesson from IBM's experience is that ninja strategies and tactics are appropriate for capturing markets but not necessarily for permanently occupying them. IBM recognized when it was time to move on to new tech territories.

How does it do it? It adapts, adjusts, and dominates.

ADAPT

Few other companies have survived and prospered as long as IBM. It's no wonder that IBM is usually seen as the "dark suit" company, without humor or a dynamic corporate structure. No Ping-Pong tables or beanbag chairs for IBM employees! But make no mistake, IBM is one of the most dynamic companies in human history. It has thrived as such by always being ahead of the next wave of technology. It helps that IBM has created a few of these waves by itself, as it did with one of its first products, the employee punch card. But even when it hasn't, it has possessed that ninja ability to adapt to changing circumstances. For instance, IBM has not necessarily been known as an "Internet company," per se, but as the Internet Age ramped up, big corporations weren't convinced their future was online. After all, it was a place for Silicon Valley geeks and too-good-to-be-true business ideas. IBM, however, showed that Big Business had a big stake in the Internet and used its strengths in mainframes and transactions to create a strategy called e-business. It was a phenomenal success.

ADJUST

When confronted with an insurmountable wall, find a way around the wall. As mentioned above, IBM came out with a home computer known as the PCjr. It was a flop. IBM adjusted and went back to its old marketing strategy. IBM also tried to get into the office-support game with its introduction of IBM OfficeVision—a kind of precursor to Microsoft Office. But the platform wasn't made to be PC-compatible, and OfficeVision never caught on.[4] Instead, IBM took the ninja route and adjusted by acquiring Lotus Software in 1995 and obtaining their more popular office suite Lotus Notes.

DOMINATE

In mainstream computer lore, Microsoft and Apple get all the glory. It's well deserved. But behind it all—working the levers, plodding along, innovating like a company possessed—there's been IBM. Rarely in its history has IBM entered a market where it's failed completely. The more common outcome is that IBM dominates, or at the very least runs even with, the rest. That's because IBM's strategy is the strategy of the ninja: driving for victory.

IBM proves that a company with a strong personality can innovate, change, and prosper. "Big Blue" has done this all while retaining its identity, conviction, and presence. IBM is certainly the long-term ninja company.

To Be the Best, Be Number One

THE INSPIRATION FOR MY PREVIOUS BOOK, *The Comeback*, STRUCK me while I attended a dinner. As I sat next to a particularly arrogant Chinese government official, we got to talking. He put his thumb in the air and raised it upward as he said, "China going up." I nodded. He then pointed his thumb downward, pushing it toward the floor, and said, "U.S. going down." I was so mad I wrote a book about it—not so much about the "China going up" part, but more so the "U.S. going down" part.

And so *The Comeback* was born. The book succeeded beyond my hopes and most of its major suggestions have been or are likely to be adopted by the U.S. Congress. And while this book is not meant to be another treatise on policy, I will examine certain governments—federal, state, and foreign—throughout these pages to explore whether their innovation and growth strategies rise to the level of the ninja.

I'll be blunt: China is a ninja, and it deserves a mention in this chapter on victory. Indeed, China has many flaws, some of them potentially fatal. Its biggest flaw is also its greatest asset. As a one-party authoritarian state, China's leaders can set a directive and expect it to be enacted. Many U.S. commentators, most notably *New York Times* columnist and bestselling author Thomas Friedman, have expressed admiration for this great competitive advantage.

In a famous column on September 8, 2009, Friedman wrote in the *Times*:

> *One-party autocracy certainly has its drawbacks. But when it is led by a reasonably enlightened group of people, as China is today, it can also have great advantages. That one party can just impose the politically difficult but critically important policies needed to move a society forward in the 21st century. It is not an accident that China is committed to overtaking us in electric cars, solar power,*

energy efficiency, batteries, nuclear power and wind power. China's leaders understand that in a world of exploding populations and rising emerging-market middle classes, demand for clean power and energy efficiency is going to soar. Beijing wants to make sure that it owns that industry and is ordering the policies to do that, including boosting gasoline prices, from the top down.[5]

Friedman was both praised for his candor and pilloried for his admiration of an autocratic state. I think there's a middle way. We can pat ourselves on the back all we want for our democratic, free-market system as the best humanity has ever devised. It is. And we can further criticize China for being an oppressive regime that stifles information, exploits its workers, manipulates its currency, and steals patents like candy from a baby. It does all that too. In his bestselling book, *The World Is Flat*, Friedman even wrote a chapter called "China for a Day (But Not for Two)." Let me go on the record and say, "Not even for a second." Freedom and democracy are too precious to hand over for even that brief period.

But that does nothing to change the fact that China *is* in the process of overtaking us, and it is doing so in no small measure because it doesn't have anything resembling our democratic processes. More, it's even harder to refute the notion that our constitutional republic, as currently arranged, has not been generating the necessary policies to help America maintain its economic superiority. In other words, if we're going to do anything about it, then let's stop kidding ourselves. China is getting it done, and we're not. That's not a judgment on either political system; it's simply a statement of fact.

A 2012 survey of global executives from KPMG found that 45 percent of respondents, a plurality, named China the next technology innovation center in the world.[6] Indeed, in 2006, China's government announced a mandate to shift its economy from the world's manufacturer to the world's innovator. Known as the "Indigenous Innovation" plan, it calls for China to become a global innovation leader by 2020. A 2009 analysis from the U.S. Chamber of Commerce summarizes the plan:

> *The result is an indigenous innovation political and economic campaign that amounts to an all-hands-on-deck call to action for the Chinese nation to roll up its sleeves and complete the mission of catching up and even surpassing the West in science and technology that began 200 years ago when foreigners with modern weaponry and transportation technology came calling as the Chinese dynastic system was dissipating.[7]*

China has made strides already, and every time I visit China I am overwhelmed by its investment in new buildings and infrastructure. They have new bridges, airports, convention facilities, and roads. They use fiat, willpower, and a captive press to push through these projects from launch to completion.

However, while China's successes in manufacturing and building are notable, they are a far cry from leading the world in creation and innovation. Innovation requires the ability to challenge the status quo. But unlike Americans, the Chinese have not been taught or encouraged to do this and they are just now catching on.

Innovation has been America's strength for many reasons: our "can-do" attitude; a free-market system that rewards savvy risk-takers; an educational system that encourages questions rather than rote learning; our First Amendment, which promotes different views without government censorship; our heterogeneous society; and our willingness to treat failure as a learning experience rather than a badge of dishonor.

By contrast, the Chinese have a decades-long tradition of copying others, do not respect intellectual property or have a culture that even recognizes it, have a bias toward conformism, and have a government that quells dissent. Given these vast cultural differences, it may be a stretch for the Chinese to shift from manufacturing to innovation.

But they are certainly trying and their strategy is respectable.

One little-noticed strategy is a formal ten-year plan (2010–2020) backed by the Chinese government to seduce the best and brightest Chinese who have moved abroad to return.[8] Headed by Dr. Wang Huiyao, vice chairman of the China International Economic Cooperation Society, part of the China Ministry of Commerce, this program directly contacts accomplished U.S.-trained Ph.D.s, many U.S. citizens, and seeks to entice them to return to China with promises of huge compensation (one person told me millions of dollars per person) and great power, including top leadership positions in Chinese universities and companies.

On a panel I shared with Dr. Huiyao in 2011, he said China had sent one hundred million students to study abroad over the past thirty years. He described his country's national talent development plan to propel China as a world leader. But Dr. Huiyao also lamented that the Chinese are not world-class innovators, a result of an education system that focuses heavily on training students to take tests rather than to be innovators. To compensate for this, he described a plan for ten thousand Chinese students to train at U.S. universities to learn the basics of innovative thinking (and open Chinese universities to a like number of Americans).

While the Chinese have developed and are executing a plan to regain their most highly educated citizens, Americans are assisting their efforts by denying legal status to the thousands of Chinese who earn Ph.D.s from U.S. universities. Once we award the degrees, we have no plan for keeping the recipients here. Instead, we send them—and thousands of other foreign-born Ph.D. graduates—to the back of the visa line. This is a terrible strategy. We are essentially training our competition. The U.S. embassy in China reports that in 2012, some 160,000 Chinese students were studying in the United States. China and other rising nations know they can't compete with America in terms of education excellence. So they smartly "offshore" that responsibility to us, then devise creative ways to lure their citizens back home.

China also is forming partnerships with foreign innovation centers as a way of getting access to cutting-edge technology. An example of this strategy in action is the January 2012 announcement that Hisense, a huge electronics company based in China, is working with MIT's Media Lab—the first alliance between the MIT Media Lab and a Chinese company—on talent training and project cooperation regarding smart technology, artificial intelligence, and human-computer dialogue.[9]

Finally, as the U.S. Chamber of Commerce study examines in detail, China is building its innovation engine through a mixture of government mandates to encourage more domestic innovation. The downside to this is that the country is increasingly protectionist and hostile to foreign investment and business. The Chamber of Commerce study notes, "Indigenous innovation seems to be a policy borne [sic] as much of China's fear of foreign domination as China's pride in its great accomplishments and desire to be a leader in the rules-based international system."

But China clearly has a ninja strategy for victory, and that strategy is working. In 2010, China surpassed Japan as the world's second-largest economy. And with growth rates averaging 10 percent a year, China is the fastest-growing economy in the world, and could overtake the United States by 2020 according to some estimates.[10] China is the largest exporter and the second-largest importer in the world, as well. With so many good things happening for China, what could possibly go wrong?

Well, there was that flaw I mentioned. . . . As a one-party state, China has a fatal weakness, one that pops its head up from time to time. We saw it at Tiananmen Square in 1989. We've seen it occasionally in the years since, though China's leaders would likely never do something as dramatic as rolling tanks through a crowd again. Well, unless that was the only option left. And that's what I'm getting at: As much as China is pursuing a ninja strategy of economic domination, it is *secondary* to the regime's survival. The Chinese Communist Party will never allow economic superiority to get in the way of its staying in power. That's just the way authoritarian regimes operate. China's leaders believe they have unlocked a system that's better than free-market economics, and that's authoritarian economics. I'm not so sure the Chinese people agree. All the wealth in the world cannot overcome a human being's innate desire to be free. The more wealth that flows in, the hotter that desire will burn.

Nevertheless, I am not judging China's political system, only whether their national economic strategy is driving toward victory. As with the previous examples in this chapter, a drive for victory—completion of your mission—is a defining characteristic of these individuals and enterprises. They do not settle for second best and if they fail, they learn from that failure to succeed elsewhere. With China, the lesson for the United States is simple: Our strategy should be victory as well. We should set as our foremost goal the continued superiority of the U.S. economy. That we have other goals at the present moment is obvious by our inability to rise above recent troubles.

And if we do make victory our highest goal, then there is no way China can match or exceed us. In the end, as history has shown, an authoritarian economy cannot compete with one premised on the unique power of free markets, free minds, and free enterprise. It's time to unleash that power once more. It's time for America to become a ninja.

CHAPTER TWO

YOUR STRIKE FORCE

Get Out of the Garage

WHAT IS IT WITH GARAGES AND GREAT START-UPS? I'M FAIRLY CON-
fident that I failed to create an innovative product because the only
thing I have ever kept in my garage is my car, which, unfortunately,
I have always used to drive to an office.

A brief history of garages and great start-ups: In 1938, David
Packard and William Hewlett started their company in Palo Alto,
California, in their garage. In 1976, Steve Jobs and Steve Wozniak
started Apple in Jobs's parents' garage in Los Altos, California.
In 1998, Larry Page and Sergey Brin started Google in a Menlo
Park, California, garage owned by a certain Susan Wojcicki (they
helped pay her mortgage—she's now a senior VP). In 1994, Jeff
Bezos, seeing an opportunity in online retail, started Amazon in
his garage in Bellevue, Washington. Even Walt Disney, in 1923,

started doodling in his uncle Robert's garage in North Hollywood, California.

And since all of them became phenomenal successes, then clearly the secret to building a billion-dollar empire is to start in a garage—and preferably one in California, or at least on the West Coast.

At least that's the romantic ideal. It's also wholly absurd. I'm certain that my list above is not exhaustive. I could probably find a few more examples of great start-ups beginning in a garage. But it would be a useless endeavor. Instead, what I would find is that 99 percent of successful start-ups and technologies definitely did not begin life in a garage.

But with successes like Apple and Google, and with movies like *The Social Network*, a fictional account of the beginnings of Facebook, we cling to these romantic myths as if they are the only ways to start and grow a successful business: isolated, maybe a friend or two, building your future empire from *David Copperfield* beginnings. I think this is a disservice to America's innovation culture, because it casts these endeavors in a romantic narrative that simply isn't true.

What does any of this have to do with ninja innovation? Just this: The modern-day myth of the ninja is of a lone warrior, operating in stealth, cut off from all allies—the heroic loner. But the historical record shows otherwise. Ninjas of feudal Japan operated in teams, usually relatively small and targeted to a well-defined mission, much like today's Navy SEALs. The only way they could accomplish their mission was to work together, with each team member knowing his specific role and how it fit within the total team effort. A ninja all alone might have been a very deadly person, but without his team he wasn't going to get very far.

A great example of this is, again, Thomas Edison—perhaps America's first Garage Guru. As most schoolchildren could once have told you, the "Wizard of Menlo Park"—New Jersey, that is, as opposed to Google's Menlo Park, California—had barely any schooling, attending elementary school classes for just three months. He also wasn't a trained scientist; he was a mechanic who was simply curious—as well as a genius. This is all true.

But there's this perception of Edison as being a one-man show. The man with 1,093 U.S. patents to his name was surely one of the most brilliant minds of all time, but he didn't do it alone. In a wonderful essay in *Slate*, Eric D. Isaacs writes:

> *It's awe-inspiring to think of Edison sitting alone at his workbench in Menlo Park, N.J., patiently testing fiber after fiber, hour after hour, day after day. It's also patently untrue. In fact, Edison was leading the world's first large-scale research and development laboratory, a highly organized, multipurpose facility staffed by a 40-person team of scientists and technicians.*
>
> *After the light bulb proved successful, Edison went on to build an even larger "Invention Factory" in nearby West Orange, a complex that included sophisticated research facilities and manufacturing capabilities. At its peak, it employed more than 200 scientists, machinists, craftsmen, and other workers.[1]*

But Edison liked his public image, as did the public. After his death in 1931, the *New York Times* eulogized him this way:

> *No figure so completely satisfied the popular conception of what an inventor should be. Here was a solitary genius revolutionizing the world—a genius that conquered con-*

*servatism, garlanded cities in light, and created wonders
that transcended the predictions of Utopian poets. . . .
With him passes perhaps the last of the heroic inventors
and the greatest of the line. The future probably belongs
to the corporation research laboratory, with trained engineers directed by a scientific captain.*

Isaacs notes the irony: "Edison was in fact that scientific captain, the executive director of a big, world-class laboratory." It
might bruise Edison's carefully crafted public persona, but there's
no shame in saying that this brilliant man needed help from time
to time. Lots of it.

As do all ninja innovators who have a great idea or product.
They might even have the know-how to get the operation off the
ground and start earning some revenue. And not just a seed stage or
pre-revenue company, either, but an honest-to-goodness, revenue-
producing, aspiring future IPO candidate! Many entrepreneurs
have been right at this exact moment, flying high, with a few employees but with no real regard for the fact that they're about to
crash and burn—that is, unless they realize their deficiencies and
start pulling together the right team.

That's because every aspiring-ninja start-up enterprise, sooner
or later, finds itself in no-man's-land—lost, short of supplies, survival in doubt. I've seen it happen countless times to promising
tech start-ups. The moment will come when all the individual
genius that went into *starting* the enterprise won't be enough to
keep the enterprise *going*. After some exciting initial success, when
they may have even "crossed the chasm," circumstances change for
the worse. Enterprises enter a realm where new problems multiply
and old answers no longer work.

At this point, the principals in the company—usually the founders—either lead their company out of no-man's-land or helplessly watch it suffer an agonizing decline, and perhaps death. But the true ninja innovator simply won't let that happen. They will put egos aside and do what's best to save the enterprise—even if that means stepping aside and letting others take the reins. At this point, it's about finding the right individual or individuals who can come in and work as your strike force. Their job? Help lead your enterprise out of no-man's-land.

In 1998, eBay was not exactly struggling, but it wasn't much to look at either, with revenue at just $4.7 million. Founder Pierre Omidyar wanted to take the company public but he knew he didn't have the right people for the job. He recognized he certainly wasn't the right person. Since its founding three years earlier, eBay had been run as if it was an extension of Omidyar's personality: laid-back and very un-corporate.[2] As Adam Cohen, who wrote a book on eBay, said in the *New York Times*, "Mr. Omidyar, who had a pony tail and wore Birkenstocks to the office, conducted 'management meetings' by having the full staff, from executives to clerical workers, gather in a circle, pass around a bowl of candy, and toss out ideas."

Depending on the circumstances, there's nothing necessarily wrong with that. But clearly Omidyar's vision for eBay wasn't matching its reality. You could say eBay was in entrepreneurial no-man's-land. Knowing he needed help to expand the company to match his vision, Omidyar brought on board, to use my generation's parlance, a "suit." Meg Whitman, then at Hasbro, was everything eBay wasn't. She had an M.B.A. from Harvard; she'd risen through very large corporate environments, such as Procter & Gamble and Disney; and, according to Cohen's account, she used lingo foreign to eBay like "gross margins."

But after one meeting with her, Omidyar declared Whitman "eBaysian." She got the job that very day.[3]

Immediately, Whitman began to bring on board executives from various industries[4] and put them in charge of the twenty-three new business categories she organized around eBay wares (sports, jewelry, etc.).[5] Despite her extremely un-eBay background, Whitman understood immediately what eBay was all about: its users. She ordered her senior executives to put up items on the site regularly so that they would understand the eBay experience from the user's point of view. She even sold the entire contents of her ski home on eBay.[6] When eBay went public later that year, the stock surged to $47 per share from its target price of $18 per share. Whitman had helped make Omidyar a billionaire.[7]

The rest of course is history. eBay is now an $11 billion company, and Whitman—also a billionaire—is now at HP. (Ninjas do not rest.) But eBay's impact on the economy is immeasurable. No other company, with the exception of Amazon, did as much to initiate the new era of e-commerce. Would that all have happened had Omidyar not decided that he needed some help, that he needed a strike force, led by Whitman, to take his "great idea" company from the garage to the C-suite?

The HDTV Strike Force

YET NINJALIKE STRIKE TEAMS AREN'T RESERVED JUST FOR START-ups. Many innovations and products that otherwise would never have seen the light of day required an efficient, aggressive strike team to make them successful. A great example of this is the task force that created and launched high-definition television (HDTV) in the United States. I was fortunate enough to be involved with this amazing group.

As a technology, HDTV is not as new as most consumers might believe. In fact, HDTV goes as far back as the 1970s (even earlier, but I'll spare you the technicalities), when Japan first developed an analog HDTV system called Hi-Vision. Although the technology required about twice the amount of bandwidth as an analog broadcast, it provided four times the resolution. Attempts were made to bring Hi-Vision to the United States, but the Federal Communications Commission (FCC) blocked these efforts because it required extra bandwidth, which was already too scarce. The goal was clear: The United States needed a more bandwidth-efficient system than analog if HDTV was to be a reality. Although we didn't know it at the time, we were on a mission that would kill analog TV.

So on January 22, 1988, the Advanced Television Committee (ATV) of the Electronic Industries Association (EIA) was formed. This consortium of industry leaders chose as its head the legendary Sid Topol, who, among many other accomplishments, was one of the founders of cable television and of Scientific Atlanta (now a Cisco company). Sid was a leader and a technologist. Most of all, he was an innovative thinker, always looking to push the boundaries of what was possible. He was just the man to lead our strike team.

As for my involvement, I was at the time a lawyer for the EIA who had volunteered to staff the effort. I gave Sid everything he needed, including draft position papers, agendas, minutes, and so on, and helped him reach out to every key player. I saw his mastery at running a meeting and getting all of the diverse team members to agree on a strategy: At the start of each gathering, Sid would recite what we had already agreed upon as well as the issues on which there were different views, and at the end of each meeting he would summarize the further issues about which we had just agreed. At the time, I thought these were time wasters and that Sid was treating us as if we were slow, but I realized that moving a diverse group from point A to point B is, as the saying goes, like "moving frogs in a wheelbarrow." I now understand and appreciate that Sid's regular summaries built conceptual fences to keep the group aligned and prevent resolved issues from being rehashed.

And, believe me, the EIA was wrestling with huge issues. Multiple entrenched industries and the politically sensitive federal government had to be involved as we shifted to a new and better TV service. HDTV is not like just any other technology; it's more properly understood to be a medium, like radio waves, in which various technologies can travel. In every country so far, a federal government has reserved the right to allocate scarce radio frequency spectrum among specific uses so that, for example, television, radio, wireless services, and defense communications do not interfere with each other. In turn, with specific uses and standards for specific frequencies, equipment manufacturers are able to tailor technologies for each unique use. And, at least in representative democracies like the United States, federal governments tend to be reluctant to set standards that would render obsolete users' existing products or require every user to buy a specific new type of product. For our HDTV team, our straitjacket was simple: Don't force people to trash their current TVs or force them to buy an HDTV.

Moreover, Japan's demonstration of its way to broadcast and receive HDTV seemed to limit our options: We didn't want to lag behind the Japanese or simply adopt their standard. This last point was contentious. Some said we should just adopt the Japanese model, but with slight improvements to meet U.S. FCC spectrum requirements. Others, myself included, argued that we had to develop a new standard for the unique needs of the United States. Compared to Japan, with its tight geography and highly concentrated population, our country is geographically very broad (not even counting Alaska and Hawaii), and our population is widely dispersed. In any case, what the Japanese had come up with basically used the old analog technology but added some new tricks. We knew we could do better.

The Topol committee adopted this "uniquely American" position. We joined with broadcasters in asking the FCC to look at the issue. Agreeing, the FCC quickly created the Advisory Committee on Advanced Television Service, which brought together all the interested industries, and wisely asked former FCC chairman Richard (Dick) Wiley to head it. Thus began a nearly twenty-year effort involving seemingly thousands of meetings and billions in private investment that ultimately resulted in the world's best HDTV standard. It felt like a reprise of Sputnik and the Moon Shot: Although Japan rushed out the world's first HDTV standard and the United States took longer, we got the job done better.

Part of our success was due to the fact that we were far more deliberate. In the end, the advisory committee would weigh more than twenty different proposals. We even financed a special laboratory (the Advanced Television Test Center, where I sat on the board of directors) to test the finalist systems. But toward the end of the planned testing—out of the blue—General Instrument (GI) announced a breakthrough: It had created an all-digital transmission system, unlike the analog systems we had been testing. Digital was new, different, and much, much better, technically and economically. Using the GI system, a broadcast tower could transmit millions of pulses per second that were received by TVs as video and sound, using algorithms to reduce the amount of information sent if portions of the visual scene did not change. My reaction? "Wow!"

Which is not to say that everyone applauded GI's system. In fact, it caused uproar among some other companies that had their own proposals. Some members of the broadcasting industry who sat with me on the board also objected. They said that the rules, process, and testing regimen had been agreed upon and it was too late to make changes to accommodate GI's breakthrough. It was a critical moment. I disagreed with the objectors and argued that the United States should have the best HDTV system possible. Chairman Wiley agreed that the digital breakthrough was important enough to rejigger the testing and delayed the selection process in order to create appropriate tests for digital systems so that other companies could try their hand at it. After all, we were all striving toward one goal. That we were doing this as Americans, united in common purpose, was not lost on anyone. Looking back, it was the right decision.

But the delay caused a problem. The FCC and Congress were expecting results based on the agreed-upon timetable. Fortunately, Chairman Wiley was especially adept at keeping FCC and congressional members informed without letting them interfere in our industry-centric process. Given that so many companies had invested so much money in the project, it was frankly amazing to me that we were able to delay the process and reformat the testing to accommodate a digital system. But as the sole representative of equipment makers on the ATTC board, my mantra became: the best system possible and fairness to all proponents (most of whom were members of my association, in any event). This was a consistent theme throughout the years-long endeavor. Often during testing, prototypes would break and some would say, "Too bad—

your system failed; you lost." Learning from Sid and Chairman Wiley, I always countered that the national stakes were too high to penalize a proposal based on a faulty component. This resulted in even more delays, but the United States ended up with the world's best system. As far as I know, no company felt cheated, and few disagreed with our final result.

Although I was representing all technology companies, I became known as an HDTV purist, arguing for the best possible American HDTV. Chairman Wiley, retailer Tom Campbell, HD-Net's Mark Cuban, Panasonic's Peter Fannon, CBS's Joe Flaherty, and Zenith's John Taylor—now that's a strike team!—joined me to lead the charge and insist the American standard have a wide screen, high-quality Dolby surround sound, and millions of pixels.

Nevertheless, internecine battles occurred daily. For example, Fox Broadcasting argued that simply transferring from an analog to a digital system, without the hassle of incorporating HDTV, was good enough. Although Fox revolutionized sports broadcasting with digital innovation, including football's visible line of scrimmage, for years it insisted on broadcasting only in DTV (digital TV) at 480 lines progressive, whereas other major networks broadcast in 720 lines progressive or 1080 interlaced. Eventually, Fox would join the others in moving toward an HDTV broadcast, because it did not want to be considered the low-quality network.

The seemingly steady progress came to another head when a few computer companies and film directors tried to change the transmission standard. I answered their demands with a simple proposal: Instead of choosing just one standard, why not embrace them all? In other words, every TV receiver would be capable of receiving every possible format but would transform them to display only the format the TV manufacturer selected.

The turmoil of moving forward went beyond the United States. When Japan saw our digital breakthrough, Japanese manufacturers first rushed to flood the market with their analog DTVs, but they quickly recognized that the United States was moving toward a superior digital HDTV standard and had to reverse course to stay competitive. Japanese manufacturers ended up recalling over a hundred thousand analog TVs.

The European debate was even more interesting. European leaders believed that consumers cared little about the quality of the television picture compared to the quality of the programming, so although they followed the U.S. adoption of digital, they did so at an inferior quality and without a wider screen. Whereas Japan erred in prematurely launching an inferior technology, the Europeans seemingly relied on the coercive power of government central planning to determine the commercial outcome. In my fortunate position within the most technologically advanced industry and one of the freest enterprise systems, I long ago came to recognize the futility of government-imposed dictates when I saw them.

In 1991, I traveled to Paris to speak at a conference on digital television. It was my first international business trip, but it didn't stop me from giving a passionate speech about the need to let consumers and the marketplace determine how good the TV-set quality should be. I argued that Europe should go for the best system possible, rather than the half step that had been proposed.

The European engineers and marketers in the audience agreed with me, but government bureaucrats ruled. Europe adopted a low-quality standard, which meant that the European standard for broadcast television did not change the basic square shape of the television, nor was picture quality as good as our U.S. standard.

It wasn't always easy selling HDTV to the broadcast and cable industry. I had several disappointing conversations with HBO and NBC, including one where Bob Costas said he wasn't interested. In fact, we approached NBC about broadcasting the 1996 Olympics in HDTV as a goal we could rally around, and not only were we rejected, but we also got a nasty letter threatening legal action in response.

But then there were the more forward-looking content providers. Chief among them was ESPN. When ESPN announced an HDTV channel, I immediately publicly lauded it as a turning point for HDTV. I quickly befriended the ESPN HDTV advocate, Bryan Burns, who continued to expand the ESPN HDTV offerings, providing the strategy, vision, and passion to lift HDTV to national prominence. To this day, Bryan is always looking beyond and is now championing ultra-high-definition television, pushing us even farther. Other great individuals in this arena included producer Randall Dark and Mark Cuban, who actually started the very first HD channel: HDNet.

Of course, when HDTV took off in the United States at the turn of this century, even European bureaucrats were forced to take notice. European satellite companies began broadcasting better signals, and all TV manufacturers started selling wide-screen sets to European markets. Despite little support from either European governments or broadcast stations, European consumers voted with their euros and selected full HDTV, which brought them the sound and colorful glory of their favorite sports, movies, and so much more.

A remarkably short history has proven American HDTV purists correct on an increasingly worldwide scale: Consumers overwhelmingly choose the wider screen and the better performance of HDTV over the squarer, cheaper standard-definition digital TV.

The HDTV story is one of a great strike team composed of

individuals—from both government and the private sector—who individually and collectively had a positive, worldwide impact on a technology whose importance continues decades later. On HDTV, we were the ninjas.

Even our tactics were likely similar to those used in feudal Japan. Before launching a mission, a proven leader was first selected, and then his team was deliberately assembled, with the requisite talent and experience, so that it was not only fully capable of confronting predictable obstacles, but also independent enough to deal with the inevitable unexpected ones. Moreover, when the mission was accomplished, the team disbanded, leaving its members available as needed to mount other missions.

And I'm proud to say that our HDTV team was an all-too-rare example of where government actually helped, rather than hindered, a free-market solution. Not only did government encourage a private-industry solution for the new standard, it also repackaged the spectrum for auction (netting over $20 billion for taxpayers) and set the timetable for the transition. The government also gave consumers a subsidy (wasted in my view) to ensure old TV sets could get the new signals.

So the next time you enjoy a game vividly displayed on your fifty-inch wide-screen HDTV set, know that it isn't one guy who worked on it in his garage whom you should be thanking.

Know Your Weaknesses

I ALMOST DIDN'T STUDY TAE KWON DO. I THOUGHT I WAS TOO INflexible. Since I was young, I couldn't touch my toes. I even hated sitting on the floor. Some people are naturally flexible; I am not.

I had watched enough karate movies to know that the best experts are able to do splits and kick all sorts of ways above their heads. I couldn't ever envision myself doing that. I believed I couldn't force my body to do things it regrettably was not designed to do.

But my kids were persistent. They wanted to study and learn tae kwon do. So we attended a free "sample" class. I found the workout vigorous and the stretching difficult, but not terrible, so we returned for another free session.

At the end of the second class, I sat down with the school manager, who explained the program. I admitted that I had little physical flexibility and expected not to get very far in tae kwon do. She assured me that flexibility was not critical for tae kwon do. That helped, but perhaps not as much as her offer to give me a special family rate.

Thus, I signed up and began my personal quest to become more "flexible."

I found the vigorous exercise of each class difficult but doable. The katas were more of a challenge, but my sons helped me get through it. It was a new and enlightening experience to look to my sons for expertise; perhaps this contributed to their self-confidence. At ages four and five they were excellent teachers. They were also better than me at video games—forever destroying the myth that their father knew everything.

But they could not teach me flexibility. Although I had a decent straight-ahead snap kick, I struggled with side kicks, which went higher than my waist, and I would overcompensate by bending my head to the floor when I kicked. Try it and you'll know what I'm talking about.

I also tried various stretching exercises, but they barely moved the needle. In desperation, I started using a special device that would forcibly spread my legs. Each day I would put a few more clicks on it and spread my legs farther. I worked on this intensely over a period of several weeks. For days I was in physical pain. Eventually, I saw my doctor, who ordered me to stop this ridiculous routine, and so I did. The pain went away.

Instead, I worked on improving my flexibility. I worked on strength. I worked on lowering my body during each side kick so the kick could go higher.

Many advice books tell you to work on your weaknesses. Certain weaknesses? Sure. But I discovered that this is not a hard-and-fast rule. There are simply some things that you just aren't good at. Tae kwon do taught me that it is just better to accept that fact. I learned my weakness and accepted it. I would not be Bruce Lee.

It's the same thing with developing your employees. I used to do annual employee reviews and would spend much of the review time discussing employees' weaknesses and ways to turn their weaknesses into strengths. But rarely did those weaknesses improve. After trying this unsuccessfully for years, I've concluded that it's best to focus on how strengths can be used to compensate for weaknesses. I also urge employees to hire people who can compensate for their supervisor's weaknesses and complement their strengths.

I bring this up because a major component in putting together a successful team is to know your own weaknesses. It rarely, if ever, helps to say, "I'll just improve on my weaknesses." In some cases that might be true. But more often than not, those weaknesses won't improve, and your enterprise will suffer the result. Think back to eBay. Omidyar could have said that he would just learn how to bring a company to an IPO. Read some books, perhaps tighten up the office environment, get rid of the candy jar. But Omidyar knew that wouldn't work. If he wanted eBay to succeed, he had to compensate for his own weaknesses by bringing in Meg Whitman. Or think of my HDTV experience. We could have rushed out of the gate like the Japanese, or initiated a government-controlled, top-down approach, like the Europeans. Instead, we formed a team that allowed each member from industry and government to provide his own strengths, while each of us compensated for the others' weaknesses—just as a ninja strike team should.

CHAPTER THREE

IN WAR, RISK IS
UNAVOIDABLE

A POPULAR USE OF THE NINJA WAS TO INFILTRATE INTO WELL-defended castles or cities against which a frontal assault would have been useless or suicidal. A ninja would sneak in using a variety of methods (none involving a giant wooden horse) and engage in activities ranging from scouting the defenses and probing for weaknesses to setting fires to distract the defenders before an attack was launched.

These missions were obviously dangerous. No culture has ever looked kindly on enemy spies. Any ninja who managed to escape detection long enough to start a conflagration would soon find himself in the middle of a burning fortress that was under assault.

These warriors knew the risks they were taking, but they also knew what the potential rewards would be, for both themselves as individuals and for the cause they served. The understanding of this concept, combined with the ability to take big risks, underscores the definition of a modern ninja innovator.

Why Risk Matters

THINGS CHANGE, EVEN IF WE DON'T WANT THEM TO. NO COMPANY operating in the free market can be successful in perpetuity by delivering the same products with the same marketing and the same margins. While it is easy and natural to crave consistency and avoid risk, the changing nature of life and our environment requires us to change, to adapt, and to take chances in order to survive.

Taking risks does not mean being reckless or random. It means exploring options, assessing likelihoods, and making rational decisions. It means seeking feedback, actively listening, and weighing the consequences of any decision *before* pulling the trigger. A ninja should be self-aware and emotionally intelligent to understand these risks.

Americans actually benefit from a considerably lenient culture toward failure. As I wrote in the introduction, some of the most successful Americans in war, politics, and business were also spectacular failures. This doesn't mean their contemporary opponents were forgiving. Indeed, there were several efforts to replace Washington as commander in chief of the Continental Army, just as Lincoln had his many detractors. Rather, the larger American public accepts failure much better than other countries. Other cultures, particularly Asian societies, have a strong tradition of avoiding shame. To bring shame upon oneself (through failure) is perhaps the worst thing that one can do. Indeed, avoiding shame is closely aligned with avoiding risk.

But in the United States, we value failure and call it experience. If you never fail at something, you are not risking enough. Few cultures are as accepting of failure and willing to view it as a positive. But, I would argue, the American view of failure is one of the principal traits that has led to our historical innovation dominance.

As for me, I am passionate about the value of failure. When I consider my life experiences, both professionally and personally, I can see that I have always learned the most when I have failed. It is now built into my DNA to evaluate any failure for "lessons learned." Failures improve me. Successes are nice, but I fear too many victories without an intermittent loss can overinflate my ego and make me less likely to listen, or make me arrogant.

Not that I desire failure. In truth, I hate it as much as the next person. I try to avoid it. But I also don't let it paralyze me. However, once I fall short, I give myself little time for wallowing in self-pity. I analyze what I have learned and I look toward the future. Early in my career, a human resources professional described me as "willing to take risks, but doing the research first." I still think this is an apt description.

Each of us has an ego and a reputation. We put these at risk every time we try something new. I imagine plenty of good ideas are never acted on because we fear failure and value our reputations. Public failure is scary. But part of being a ninja is understanding that with mistakes come experience, so that even in failure, we gain.

People also tend to take fewer risks as they get older. Some even believe that starting a business is for the young. Indeed, Apple, Facebook, Google, and Microsoft were among the thousands of successful businesses started by founders under thirty. Earlier in life you may be risking only your own economic well-being. Later in life you may have a family and responsibilities to others. You also may have a certain lifestyle you do not want to risk.

In certain areas, I have taken on more risks as I have become older. For important issues, I have taken risks and become outspoken. For example, I feel strongly that our country is unfairly stealing from the next generation. I have spoken out loudly on this issue—even if it means criticizing prevailing leaders from both major American political parties. Yet, my risky forcefulness has paid dividends. I have found that while speaking truth to power may not endear you to it, it does gain you respect. My willingness to take a risk and speak out has landed me on various lists of influential policy people.

Now, this isn't an invitation to put it all on red in a Las Vegas casino or buy an obscure penny stock. It is a call for you to explore and learn your own behavior and risk tolerance, and be willing to take a risk for something worthwhile and view failure, if that's the result, as a learning experience.

Intel Inside

IF ONE THING DISTINGUISHES INTEL'S INNOVATIVE NINJA THINKING, it is their 1990s strategy of branding a semiconductor chip as a valuable feature that consumers would look for when they purchased a computer. The campaign's two decades of ubiquity make us forget this now, but at the time it was an incredibly novel approach to marketing. People bought computers because of the software, the specs, or a friend's recommendation. Who cared about who made some tiny chip inside the box that you couldn't even see?

But with the proliferation of PCs, and with consumers at a loss in trying to figure out what made one better than the other, Intel saw an opportunity, and so it took a major risk. Intel's leadership was convinced this was the way to grow market share, however, and the company invested hundreds of millions of dollars in the effort.

I first became aware of what they were doing in 1993, when I met with Intel executives to discuss their International CES participation. They described a plan for a remarkable exhibit at the show that year, a centerpiece of their effort to shift the image of Intel from that of a chip company to that of a producer of a coveted, brand-name product that stood for performance. The display Intel was planning was so elaborate that we actually had to work out a deal so they could occupy the same exhibit space they would be using at another show taking place a month and a half earlier.

The booth was a hit, and for several years Intel built on its success at CES by rolling out ever-larger, more elaborate and immersive exhibits showcasing the company's cutting-edge technology. I will never forget being inside the "Intel experience" and feeling like I was touching the future. By using an amazing multimedia cascading exhibit, Intel managed to convey to me and others that Intel was way ahead of almost everyone else. You can't get the same impression by reading a two-dimensional ad in a magazine or even watching a television commercial. Intel wisely used the live experience to change the very perception of its company, its products, and its importance.

Intel also cleverly used its CEO keynote and marketing around the show, including signage, publications, and live events, to ensure that every CES visitor knew about "Intel Inside." Soon, consumers looked for that label before buying a computer, much in the same way that they look for the American Dental Association Seal of Acceptance when shopping for a toothpaste. By marketing itself in that way, Intel transformed into a brand known to millions of otherwise technology-illiterate consumers. Those consumers might not have known a motherboard from a mainframe, but they had "Intel Inside."

The Intel marketing campaign taught me three things:

First, a clever company can create something out of almost nothing by thinking outside the box. Intel turned a chip into a brand and that brand into billions in added sales.

Second, Intel's success demonstrated the power of trade shows to create an indelible live, interactive marketing experience. Intel captured the critical influencers who attended the trade show, including the media, retailers, and Wall Street. Intel's marketing wizardry impressed upon them that the company was becoming not just another chip fabricator, but a brand in its own right.

Finally, Intel taught me the value of a visible CEO in enhancing and transforming the image of a company. Almost every other year, Intel CEO Craig Barrett delivered a keynote address at CES. (His successor, Paul Otellini, delivered the CES keynote in 2012.) Barrett used every second of his allotted sixty minutes to demonstrate how Intel "got" the future. More, I had the sense that Intel was at the center of the product offerings of so many other companies. The CEO would not only convey factual information about the company and its products, but also leave every audience member impressed by the importance and future of Intel.

Risk-Taking Starts with Belief

ONE OF THE BIGGEST REASONS INTEL WAS WILLING TO PUT ITS brand directly in front of consumers is because the company believed in itself. The company's executives knew Intel had the best technology and manufacturing processes, and could back up its tremendous ad spend with performance.

Leaders of companies that believe in the future have a passion for what they are doing. They convey excitement, interest, desire, and a level of intensity that transcends the common, blasé tone that pervades much of society.

In my many years in the industry, I have had the chance to meet and work with many passionate leaders. I've been blown away by people like John Chambers of Cisco. Watching John work a room is a delight. He asks penetrating questions, feeds off the energy in the room, and shares his enthusiasm, interest, and excitement. I've seen him give speeches, brimming with energy as he dominates the stage, bubbling with enthusiasm for his presentation, and using his entire body to emphasize his main points. His passion moves his audience and transforms listeners into believers.

Amazon's Jeff Bezos (see chapter 5) is another CEO who believes in his company and leads with passion. Jeff is not physically imposing; in fact, he wouldn't be chosen first for any sports team. But what he lacks in physical presence, he makes up for with a confident presentation style that combines a self-effacing nature with a conviction that his path is the way to go, as unlikely as that might seem to whomever is in the room.

I will never forget two of Jeff's talks to our group. In the mid-1990s—long before most people were even online—Jeff already knew how he would change the world. He described how Amazon would become a seller of choice of all products, starting with books. At the time, I recall believing him, based on a persuasive argument that was compellingly delivered.

Fast-forward a dozen years later when I again had a chance to hear him speak. Talking to our group in California, Jeff laid out a plan for a new type of book that could be read electronically. I recall my colleagues and I squirming a bit at the idea. After all, this was a guy who'd changed the world by delivering paper books—and this would be a totally new area for Amazon. The company had a massive infrastructure in place for shipping physical goods; now it would need to spin up a similar virtual warehouse for online content.

Of course, Jeff was right. I look forward to his next big idea. With his track record, experience, and passion, I'm sure it will be a winner.

It does take more than passion to be a worthy leader. For several years, Countrywide Financial's Angelo Mozilo was everywhere touting his company and its mission of allowing every American to buy a home. Every time he would appear, my wife would tell me she didn't trust him. Turns out she was right. Countrywide made millions of loans to unqualified home buyers and helped cause the real estate crash that almost brought down the world's economy.

France's "Defensive Strategy"

MOVING ACROSS THE OCEAN, WE CAN LOOK AT A CULTURE THAT does not value risk. France is a strong, rich country, with the world's fifth-largest economy as measured by GDP. More than eighty million annual visitors make it the most-traveled-to country in the world.

France's core strength is that it is a sensuous country. It excels in grabbing, shaping, and enhancing each of the five senses. Its wonderful wines and food cooked magnificently thrill the palate.

Its perfumes, gardens, luxury goods, and cuisine delight our olfactory senses. The French language, with its curves and smoothness, seduces the ear. Dozens of words we use in cooking, food, and restaurants are actually French words.

France is also visually enchanting. It's not only the couture fashions and thousands of delightful shops that attract the beautiful, the well dressed, and the elegant. It's also the bustling but sophisticated French streets and the entire French culture. The museums, especially the Louvre, burst with incredible classics of art. The magnificent country contains charming towns and pastoral settings that inspire both desire and contentment. France's rich visual display is a vital part of the allure of the millions who visit France.

So as a sensual culture with income that derives significantly from food, fashion, and foreigners, how does France innovate and grow its country? It has a strategy based on its strengths, but its efforts to build up barriers to protect its language, culture, government, and increasingly dysfunctional workforce threaten its success.

Like many countries, France learned the wrong lessons from the First World War. As war with Nazi Germany approached, France's generals expected that conflict to look much like the last one had—a defensive struggle where gains would be measured in feet and a well-fortified army could hold out for years until help arrived. France applied these lessons toward the creation of the Maginot Line, a massive system of interconnected fortifications along its border with Germany.

The French strategy was defensive. It built an impenetrable wall and dared the Germans to cross it. The Germans took one look at the Maginot Line and said to themselves, "The French are right. It is impenetrable. So we'll just go around it."

That's because the Germans had learned entirely different lessons from the First World War, and the innovative new doctrine of blitzkrieg—or lightning war—combined with an attack through Belgium exposed the Maginot Line as being entirely unsuited for the modern way of war.

Blessed with an amazingly rich cultural heritage, the French are again trying to erect a Maginot Line of sorts to protect what they hold most dear. Unfortunately for the French, there is little hope that this new wall will work any better than the last one.

Start with language. For hundreds of years, the French language was the tongue of European diplomacy. A far-flung colonial empire expanded the language's reach to the rest of the world. Even today, it remains an official language in twenty-nine countries, with a majority of the world's French speakers living in Africa.

Yet French long ago ceded its dominance to English, which has become the global language of business, science, music, movies, and air traffic. Close behind are the huge number of speakers of the various dialects of Chinese, Spanish, Hindi-Urdu, and Arabic.

The French language has lost its prominence, but the French stubbornly defend it from outside influence. The French agency L'Académie Française, first founded in 1635, is the official custodian of the French language. It is entrusted with writing the official French dictionary and even though its dictums are not legally binding, the very existence of the body reflects the disturbing French habit of playing defense. For instance, in 2011, the agency set up a website that would list "blacklisted" words that had crept into common French usage. A report in the *Telegraph* (UK) said:

> *Agnès Oster, secretary of the body's dictionary commission, told* The Daily Telegraph *that more English terms would be added to its online blacklist every month.*

> *November's additions will include the franglais term "supporter" to mean "support" (a team, for example). It suggests replacing it by "soutenir" or "encourager."*
>
> *It will also urge French-speakers to drop Anglicised superlatives like "top," "must," or "hyper" using instead proper French terms like "incomparable," "très bien," or "inégalable."*
>
> *It also hopes to wean them off the cinema term "casting" and replace it with "passer une audition."[1]*

Unwilling to change, the French view this as a defense of their cultural heritage. Indeed, there is considerable support in the United States to make English the official language. But this is more symbolic than anything else. Most Americans would laugh at any government attempt to stop the natural evolution of language, which, like biological evolution, is impossible to prevent.

Here again we can draw a distinction between the French people (plenty of ninjas) and their government (not so much). While the French government stubbornly sticks to its "protect French" mantra, the more ninja-esque business community has figured out that it must be flexible.

As the Chinese economy has boomed, more and more Chinese tourists have come to see the amazing sights of France. Near the world-famous Louvre in Paris sits the famous French store the Galeries Lafayette. Spend some time sitting nearby and you'll quickly notice bus after bus full of Chinese tourists arriving to check out the wares. The store has responded to this demand by hiring employees who speak Chinese. According to store personnel, half of their four hundred sales employees now speak Chinese, to better serve the tourists who now make up the largest segment of its customer base. (Americans rank fifth behind Russians, other Europeans, and Arabs.) The average Chinese tourist at the store spends over €5,000 (nearly $7,000)!

Other Parisian retailers also are scooping up Chinese cash. On a recent trip to Paris, my wife witnessed a line outside the Louis Vuitton store near the Galeries Lafayette. The store has started using a velvet rope to manage the mostly Chinese crowds wanting to shop there.

The lesson is clear: Businesses will adapt, even in France, *if the government doesn't get in the way*. Some important changes France should make:

First, abandon the extreme protection of the French language. English is the new number one language. A monolingual populace will hurt French primacy as a destination for tourism and visitors.

Second, stop protecting French culture by mandating the domestic content of television and radio. France has isolated itself and hindered growth by walling off international competitors and restricting outside access to French movies and music. The country has uniquely enacted strange laws protecting "moral rights" of content creators—which allow a creator to always have some say over his work regardless of who owns it—and its "three strikes" law disenfranchises consumers of content. These restrictions prevent many French artists from breaking into global awareness. It also stifles competition that would improve the domestic film and music sectors in France and help turn content creation into a major national export.

Third, do something about the antiquated, job-killing labor laws. The French, like the rest of the European Union (EU), suffer from a lack of work commitment, as reflected in their laws setting mandatory minimum vacation days. EU laws require a minimum of twenty paid days of vacation per year, not including national holidays, and EU-mandated weekends, breaks, and time off for night workers. More, the EU in 2012 even ruled that workers who get sick on vacation days are entitled to an equal amount of additional days off.

But France goes farther than these already generous mandates. The workweek, by law, is set at thirty-five hours per week. Anyone who works thirty-nine hours per week gets an additional ten days of paid leave each year. With such generous holidays and mandatory leave benefits, it is not clear why French workers so frequently strike. Almost every time I have been in France, some major strike affected my travel plans. This work attitude eventually will also hurt France's core strength of attracting visitors. Most travelers do not have great flexibility in arriving and departing and cannot be held up for days due to striking workers. People understand weather issues, but they resent strikes.

These types of mandates, with the guarantee of lifetime employment and laws that make it very expensive for companies to fire employees once hired, make France a less desirable location for business investment. So unless France changes its attitude toward employers, it will remain a wonderful place to visit, but not a great place to create jobs.

Back Off the Internet

THE FRENCH ARE CERTAINLY CAPABLE OF INNOVATION IN TECHNOLOGY, as proven by the Concorde, Renault, and especially the country's successful nuclear program. More than three-fourths of the energy produced in France is derived from nuclear power, and the country is at the forefront in developing next-generation reactor designs and the recycling of spent fuel.[2]

Despite these successes, overall France has a terrible record of trying to anticipate technology and pick winners and losers. (In fact, every country does; some just try it more often than others.) The government of François Mitterrand nationalized the Thomson-Brandt consumer electronics company in 1982.[3] Five years later, the company acquired the already declining American company RCA for approximately $1.1 billion in cash and other considerations.[4] Thomson then sank another $1 billion into a turnaround plan.[5]

In the face of continuing losses, France looked to China for help, forming a joint venture with TCL in 2004. Just three years later, after TCL had taken a $680 million loss on its new European operations, TCL closed things down.[6] Thomson ended up selling off the RCA audio/video and accessories brand for only $50 million.[7]

Unfortunately, the French government has not learned from past mistakes. In May 2011, then-president Nicolas Sarkozy addressed more than eight hundred top technology executives assembled at the eG8 conference in Paris regarding his proposals for increased government involvement and control over the Internet.

Sarkozy began by describing the Internet in grand terms as "a new form of civilization" and "the third globalization."[8] However, he then made clear his view that this new civilization could use a lot more civilizing. He insisted that governments must dramatically increase regulation of the Internet to protect intellectual property, children, privacy, and security, and to ward off monopolies. He wanted to push this innovation-smothering approach onto other countries, and he was asking this crowd for support.

I was at that summit and can report that these remarks did not go over well among an audience that included influential business leaders like Jeff Bezos of Amazon, Eric Schmidt of Google, and Mark Zuckerberg of Facebook. One questioner from the United States said the Internet was the "eighth continent" and governments should follow the example set by the Hippocratic Oath and "first do no harm." The comment elicited strong applause and forced Sarkozy to defend and repeat his vision of why governments must regulate the Internet.

We expect to see calls for censorship from regimes that govern without the consent of their people. It's shocking, however, to see a democratically elected leader calling for worldwide regulation of the Internet. And it was downright thrilling to see the American innovation community pushing back that day with questions, applause for the questioners, and a press conference.

The American approach—with the Internet open for innovation and a force for freedom—won the audience. The crowd recognized that the Sarkozy proposal, whatever its motivations, threatened the very openness that has fostered two decades and counting of blisteringly fast innovation.

Too much is at stake for America and the world to allow the Internet to be constricted by old government's quest to control content. As Google's Schmidt has argued, the men and women on the leading edge of technological innovation are smart enough to see the potential problems, and innovation itself will do a much more effective job at addressing potential problems than behind-the-curve government regulators could ever hope to do themselves. Our president and Congress must lead the nation and the world by resisting these calls to regulate the Internet.

In a panel discussion that took place after President Sarkozy's remarks, Schmidt and eBay CEO John Donahoe agreed that government's role is to provide citizens with Internet access, not to regulate content. Sadly, not all were convinced, as it fell to then–French finance minister and future International Monetary Fund managing director Christine Lagarde to defend Sarkozy's position, insisting that a lack of regulation would spell "chaos."

Overconfidence Is a Weakness

I INCLUDED THE FRENCH APPROACH TO THE ECONOMY IN THIS chapter on risk because it's an example of a strategy that includes no risk whatsoever. In fact, the French strategy, like the Maginot Line, is all about *containment*—keeping what they have. And the foundation of all defensive/containment strategies is a lack of confidence in your ability to compete in the marketplace. It's the siren song that few dying industries and companies can ignore. But it's also a surefire way to lose the very thing you're trying to save. For all its delights, France is in danger of becoming a theme park for tourists—much like Italy is already: a tourist attraction where visitors behold a once-great society.

But this is not to say that all risk is good. Taking foolish chances, for example, can put you in just as much trouble as taking no risk at all. Rather, the ninja company takes *calculated risks.* For every time a company bet the farm and won, there are hundreds of examples of failure. And the springboard for taking that foolish risk is usually overconfidence. A ninja might have great confidence in his skills. He might even be a bit cocky. But there's a sea of difference between the ninja who tries to take on an army because he believes he's invincible and the ninja who chooses to flee to fight another day.

In a country like ours that rewards risk and accepts failures, tech entrepreneurs are almost universally celebrated. Among Democrats and Republicans alike, you're probably not going to hear, "Well, the problem is that we have too many tech entrepreneurs," on *Meet the Press*. Even in this world of gross economic ignorance, it's generally recognized that entrepreneurs' enthusiasm for risk-taking is an essential driver of innovation and economic growth. And tech entrepreneurs wear their risk-taking credentials as badges of honor.

That's why it's sobering to imagine that what really drives entrepreneurs is not a virtue but a usually fatal character flaw. A flaw known as overconfidence.

According to a 2003 scholarly paper, "Financial Contracting with Optimistic Entrepreneurs: Theory and Evidence":

> *Launching a company is a risky business, with about half of all newly created companies folding within four years of their creation. It would therefore seem that entrepreneurs are world-class risk-takers.*
>
> *However, what drives entrepreneurs to launch companies is not their attitudes toward risk but their perception of risk, say management scholars and psychologists. Simply put, many entrepreneurs overestimate their odds of success.*[9]

This is the conclusion of researchers Augustin Landier, an assistant professor at the University of Chicago's Booth School of Business, and David Thesmar, member of ENSAE, a prestigious French research organization. The study used data from about twenty-three thousand French companies to examine entrepreneurial optimism and its effect on capital structure and performance. The paper is directed at investors as a warning about the mind-set of most entrepreneurs, but I think it's also useful for our discussion here.

Landier and Thesmar looked at the differences between entrepreneurs and professional investors in their beliefs about new business ventures. Needless to say, there was quite a gap between their respective perceptions of risk, to the detriment of entrepreneurs.

Figuratively putting entrepreneurs on their analyst's couch, Landier and Thesmar tried to determine the characteristics of entrepreneurs most related to their overoptimism. Among their more provocative findings:

- As educational level and career experience increase, entrepreneurs are more prone to overoptimism. Because entrepreneurs must possess strong positive feelings about their ventures to put aside excellent jobs elsewhere, those who have more education and experience are likely to be more optimistic.
- Optimistic entrepreneurs aren't likely to abandon their dreams quickly or easily. Even if it becomes apparent that their business will not succeed without abandoning some portion of the business plan, adapting it, or scaling back the business, optimistic entrepreneurs' strong beliefs in their venture may prevent them from adapting to early feedback.
- Optimistic entrepreneurs believe the financial markets underestimate the potential of their ideas. Therefore, they prefer to use as much of their own and their family's wealth as possible [to finance their venture].[10]

Through painful experience—be it institutional or personal—professional investors know they will repeatedly encounter all the above (and more) in their relationships with overoptimistic entrepreneurs. It's one of the reasons that it's an article of faith among professional investors that there are more good technologies than there are good entrepreneurs or investment opportunities.

In fact, it's probably also an article of faith that professional investors' Holy Grail investment opportunity is a technology that doesn't depend at all on the foibles of imperfect humans for its success. But as long as business success depends on the skills of a management team (see chapter 2), investors will always view the brash, overconfident tech entrepreneur with the "can't lose" product with a healthy degree of skepticism.

What does all this mean to ninja innovators? For starters, it's another warning that the life you have chosen is incredibly difficult and treacherous, no matter what the size of your dreams and ambitions is. Just as having the best idea in the room isn't enough, simply taking a risk isn't enough, either. Like the ninjas of feudal Japan, whose skills were developed and perfected over years of intense study and training, today's innovator needs so much more than a product and a "damn the torpedoes" spirit to succeed.

As George S. Patton, someone who knew a lot about risk, once said, "Take calculated risks. That is quite different from being rash."

CHAPTER FOUR

PREPARE FOR BATTLE

First Is Not Always Best

TECHNOLOGY, AND ESPECIALLY CONSUMER ELECTRONICS, IS A CU-rious beast. Its primary purpose is to make our lives, our jobs, and our civilization perform more efficiently, with fewer delays from idea to execution. Technology like mobile phones, tablets, DVRs, and GPS navigation units have revolutionized our daily lives. Activities that once took a long time to do—talking on the phone, waiting for dial-up Internet, recording TV programs, or finding directions with a map—now take less time. In an ideal world, the time we save should enable us to do even more productive activities. Of course, the reality is something else entirely. As many employees have surely discovered, because information delivery and access is near instantaneous, more is expected of us than ever before. All that supposed *free time* is now consumed with even more work.

But that's a discussion for another book.

The flip side to the instant-gratification nature of our gadgets is that every one of them took *a long time* to produce. Usually years, but sometimes decades. The ninja innovators credited with either their invention or their mass distribution (not always the same person or company) spent thousands of hours planning, strategizing about, and marketing a product so that it could pass the many hurdles to hit our homes or offices. In other words, the instant gratification these gadgets engender do not serve their creators well. A ninja innovator must be patient, deliberate, disciplined, and—perhaps most of all—*fully prepared* to meet the enemies that will surely array against him in the field of battle.

I've mentioned before that inventing a product or just being first to market does not guarantee success. In fact, that's usually where your problems start. Take the first company to introduce a tablet PC—Microsoft. In 2000, Bill Gates personally introduced a tablet PC prototype, which relied heavily on handwriting-recognition software, even though he said the product would not be ready for market for another two years.[1] At the time, we must remember, digital devices that used handwriting, such as the Palm personal digital assistant (PDA), were quite the rage. It made sense to create a hybrid Palm-PC that could perform a variety of functions.

As a source at Microsoft told *PCWorld* at the time: "We don't believe users want another companion device. They want the same functionality as they get from a notebook computer."[2] A year later, Gates was even more bullish on the PC tablet: "The Tablet is a PC that is virtually without limits—and within five years I predict it will be the most popular form of PC sold in America."[3]

Over the next few years, Microsoft would partner with PC makers—Compaq and Lenovo, specifically—trying to create demand for tablet PCs. But it wasn't catching on. The devices that did come to market were expensive—upwards of $500—and with their focus on handwriting recognition and day-planning software, they were geared toward a business market. Mainstream consumers, who didn't see the value of being able to handwrite on a screen electronically, opted to continue buying laptops that met both their portability and personal-computing needs much better that Microsoft's various tablets.

Finally, in spring 2010, Microsoft canceled development on its latest tablet product, Courier, which was going to have two screens and open and close like a book. Not coincidentally, that was around the same time that Apple released the iPad, which sold more than three million units on its first day. Suddenly, it seemed that everyone was crazy about tablet PCs. So where did Microsoft go wrong?

It would take more time than we have to answer that question in full. Let me just posit that Microsoft, as ahead of its time as it was, missed the mark because the technology *that consumers wanted* just wasn't there. Or, to put it another way, Microsoft's tablets relied on handwriting-recognition software because that's what the technology allowed, but it wasn't what consumers wanted. Instead of learning what consumers wanted, which was an even smaller, sleeker, fully operational PC, Microsoft went ahead and guessed. It guessed wrong.

Apple, however, was prepared for battle. It saw Microsoft's struggle with the tablet PC. It could have concluded that it would be a waste of time to try its hand at the market, given Microsoft's problems. But Apple also saw the runaway success of its iPhone as a signal to what consumers really wanted. While Microsoft created a product that was more or less a glorified version of the Palm PDA, Apple decided to make an enhanced version of the iPhone— without the phone functionality of course.

The ninja lesson is simple: Forcing a product on the market is never a good idea. A ninja must understand the lay of the land completely if he hopes to achieve success. That takes meticulous upfront planning to understand: (1) what consumers want; (2) if the technology can meet that desire; and (3) that sometimes you just have to wait for 1 and 2 to come together. It's that last point that trips up a lot of companies. If you have the idea, and if the technology is *kind of* there, why wait for a competitor to get there first? Microsoft's adventure with the tablet PC is Exhibit A. Impatience is not a good attribute for the ninja innovator.

There Is No Quick and Easy Path

ONE OF THE UNFORTUNATE CONSEQUENCES OF OUR TECHNOLOGY revolution is that the concept of delayed gratification is ever more elusive, particularly among our children. Not so long ago children had to create their own imaginary worlds to act out. Nowadays, they simply plug in and, as the saying goes, tune out. Entertainment is no farther away than the television.

A now-famous study conducted in the late 1960s revealed the benefits of delayed gratification. In what became known as the "Stanford marshmallow experiment," four-year-olds were each given one marshmallow but told that if they waited fifteen to twenty minutes to eat the first marshmallow, they would receive a second one as well. Follow-up research found that the children who were able to delay gratification scored better on the college-entry Scholastic Aptitude Test (SAT) than the children who had been unable to resist immediately eating the single treat.[4]

The children who were able to delay gratification would have grown up to make good ninjas. The ability to put off satisfaction, which can also be called patience, is key to a ninja's success. Closely related to the discipline of patience is the maturity to use your time wisely to gather information about yourself and the world in which you're operating. To succeed, you must mentally steel yourself for the trials ahead. You must observe, study, and analyze the field.

Being married to a doctor, I know firsthand how doctors sacrifice and delay gratification their entire lives. They spend about a third of their life going through the process of learning and preparing—college, medical school, residency, internship, and fellowship—before they are established as medical specialists and can start practicing on their own. Countless hours of study, financial investment, and sacrificing socializing time all eventually pay off. Doctors—and their loved ones—understand delayed gratification.

My in-laws, both of whom are doctors, know the value of patience better than most. They suffered through World War II, the Holocaust, and Communism to become doctors in their native Poland. When they escaped to the United States, they had to learn English and retake the medical boards in the mid-1960s. They worked long hours, lived a very modest life. Now, in retirement, they receive gratification from all that they accomplished and their daughter's stellar career as a top retina surgeon.

Delayed gratification is a particularly challenging lesson to teach your children. As parents we struggle with how to raise confident, healthy children. I suggest this is best accomplished by requiring children to work for much of what they get, regardless of whether the parents are able to afford for the children not to work. This cuts against the more natural instinct of sharing your success with your children and shielding them from any harm or bad feelings. But teaching them to delay gratification and invest in their own success will prepare them for ninjahood.

Becoming a ninja is in its own way one long exercise in delayed gratification. Becoming a black belt in tae kwon do requires investing thousands of hours in rigorous exercise, practicing routines, and learning self-discipline. Show me an accomplished student of martial arts, and I'll show you someone who knows all about delayed gratification.

Similarly, show me a successful CEO, and I will show you someone who has endured, sacrificed, studied, and worked hard to achieve his or her post. Even those who inherit a family dynasty are most successful when they are required to invest in themselves before rising to prominence in the business. Professional athletes aren't the only ones who benefit from rigorous training; we all do, whether we're artists, businessmen, or even bureaucrats.

Many companies have even made delayed gratification part of their successful branding and marketing strategies. Apple is probably the best recent example of this skill. The company has found great success in essentially teasing its customers with product launches. The day after Apple unveils its latest iPhone model, tech geeks start speculating about when the next model will come out and what new features it will include. The anticipation drives some people crazy (literally, in my opinion, judging by the lines that begin outside an Apple store days before a product launch), but it also makes the reveal, when it finally comes, even more spectacular.

Apple does this very well now, but they didn't invent the idea. For decades, technology companies have employed similar strategies by waiting to unveil their newest products and concepts at the

International CES. They realize there is a significant benefit to waiting for CES, when the eyes of the world are on the industry and they can truly show off. This has been especially true with the rise of new media, where boundless websites and blogs nurture our obsessions with gadgetry and innovation.

But the concept of delayed gratification goes much farther than just delaying a product launch until the right moment. It's more than just sitting on your hands and waiting for the right time; it's about recognizing the value in taking your time and using it to gather information, learn, grow, adapt, and prepare for the next steps you'll take, whether personally or professionally. Don't think for a moment, for example, that Apple is merely teasing its loyal customers with long waits for the next model. The company, like any that employs these strategies, pays attention to the speculation and uses the time to gather information about consumers' desires and expectations. The end result is a product that meets users' needs better.

The story I shared earlier about the creation of HDTV is a great example of this concept. While the Japanese rushed to get HDTV out quickly, the United States took its time—nearly twenty years—to gather intelligence, review dozens of concepts, and analyze the ideas being proposed to bring HDTV to American viewers. And when last-minute information came in to challenge the course we'd taken, we took the time to go back to the drawing board and reevaluate our course. The Japanese may have had it first, but we had it best. In the end, they ended up recalling their analog HDTVs when the digital standard embraced by the committee proved to be superior. (The Europeans, as I noted, are a good example of waiting *too long* to move forward on an idea and failing to use the intervening time productively.)

Know Your Terrain

THE PRIMARY ROLE OF THE FIRST NINJAS OF FEUDAL JAPAN WAS TO be spies. The warlord would send ninjas to gather as much information (intelligence) as they could on the enemy to gauge strengths and weaknesses. In popular culture, the ninja employs the techniques of disguise and infiltration as a way to assassinate his opponent. But there is little evidence that proves that ninjas actually carried out assassinations. Rather, research shows ninjas developed these techniques so that they could gather intelligence. A contemporary document at the time said: "Concerning ninjas, they were said to be from Iga and Kôga, and went freely into enemy castles in secret. They observed hidden things, and were taken as being friends."[5]

Then as now, the ninja is always gathering information.

My job requires me to attend lots of lunches and dinners, where I'm usually seated with people I don't know very well. As part of my own ongoing information-gathering mission, I always challenge myself to learn something from the people with whom I am seated. It's almost a mental game I play—but one that is rewarding. Almost everyone is an expert at something, and chances are the other person is proficient in something you're not. Delving into another person's knowledge is a strategy I employ that offers me the reward of insight. For the ninja innovator, you never know when a casual conversation will lead to learning an important piece of information or about a future opportunity.

This requires real conversation, which necessitates a lot of active listening. This means not just mentally waiting to say your piece after the other person has finished speaking; it calls for thinking about what the other person is saying and asking a relevant follow-up question. One simple way to ensure you're listening actively is to restate what the person told you, to show that you were engaged and make sure you understand what they were saying. Active listening helps avoid confusion and misunderstanding and can help make any exchange more meaningful.

Doctors must be especially good at this kind of skilled information gathering. I've learned to share as much information as possible with my doctors. As experts who've invested the time in learning about the various functions (and failures) of the human body, doctors who are good intelligence gatherers will know which information is worthwhile and what is just the ramblings of a sick patient.

In my own career, I've come to learn that the same thing is true for most professions. Even if you're in a position of real or perceived authority, there are good reasons to listen more than you speak. For starters, you learn more. If you only talk, you never learn anything from other people. The old saw about having two ears and one mouth, and using them proportionally, is simply good advice.

The maxim about listening is especially true in sales. While the stereotype of a salesperson may be a fast-talking pitchman armed with an answer to every objection, the most effective salespeople are those who take the time to listen and understand what the prospect needs, what challenges they have, and how the salesperson can provide a solution to the unique situation and problems of the prospect.

The same is true in job interviews. While good interviewers are excellent listeners, an employer (speaking from experience here) will have a more favorable impression of a candidate if the conversation time is more balanced than a simple one-way question-and-answer session. Many people in power positions are more impressed if the candidate can get the interviewer talking as well. It shows inquisitiveness, initiative, and a certain amount of boldness.

For example, our better U.S. presidents, in modern times at least, have all been great listeners. As each of them works a room it becomes evident that they spend a lot more time listening than they do talking. They know how to interview for a job. The vast majority of successful political leaders are good listeners, because they have to be. Voters want to feel respected and understood.

But not all politicians care about listening. One that immediately springs to mind is a former presidential candidate who has a brilliant political mind but always feels the need to prove that he is the smartest man in the room. During every encounter I have had with him he has been content to pontificate on the issue at hand, not seeming to care much about anyone else's opinion. I have had many bad experiences with another certain congressman who also is a nonlistener—more, he is rude about it. I have testified before him several times. Most times he has been talking to someone else, reading a paper, or working on his BlackBerry—even when he was chairing the subcommittee that called me to testify. His arrogance was palpable and disconcerting to me and the other witnesses. By comparison, almost all the other members of Congress sitting on the same subcommittee would listen carefully to the testimony of both sides on an issue.

Being a good listener is just the first step to being a good conversationalist. The second step is knowing what to talk about. Eleanor Roosevelt said it well: "Great minds discuss ideas; average minds discuss events; small minds discuss people." Not every conversation can or should be about ideas. But "idea" conversations are the most provoking, meaningful, and productive. They permit insights into the other person and their thoughts. Every business, every invention, every improvement, started with an idea—and probably a discussion (or two, or ten, or a hundred) about it. Discussing ideas can catalyze growth, create energy, and establish a bond among the participants, who can leave a conversation knowing they shared something important.

Any conversation is really managed by the person asking the questions. It should never be an interrogation (unless it is), but polite questioning can reveal that person's thinking or elicit information you wouldn't have known otherwise. As the president of CEA, I do a fair bit of interviewing of leading government officials and businessmen onstage before large groups of people. I find the questions I ask at the dinner events I mentioned earlier are the same ones I ask onstage at events like the International CES.

My technique is to simply ask questions and follow up the answers with questions telescoping further into promising areas. *How's business? What keeps you awake at night? Where do you think you/your business will be next year at this time? What excites you? How did you start the business? How did you meet your spouse? How did you propose? Tell me about your family. What was your first job? Where would you invest? What are areas for growth? What advice would you give to someone starting out in your industry? What's your business philosophy?* Any answer to a question that is incomplete or expresses feelings is an opportunity for a follow-up "why" question. *Why were you frustrated? Why was it important to you? Why did you leave (the job, situation, location)?*

Notice that none of these questions can be answered with a yes or no. All require some thought and an explanation, and they are all applicable to virtually any person. It's not just about professionals, either. It's about life experiences and hobbies. By asking such questions of my professional colleagues and acquaintances, I have learned about mountain climbing, fishing, stamp collecting, dog shows, and investing.

A simple conversation with any random person is an opportunity to engage another human, learn more, and expand your knowledge base. Listen carefully to the words they use. Words make statements about a person's upbringing, culture, and values. Bad grammar is a fair indication of education and upbringing. Cursing frequently reflects a lack of control, concern, culture, or all three. Using big words when little ones would work fine tells me the person lacks confidence but wants to appear smart.

Using these seemingly simple techniques helps me learn and grow. Finding opportunities for personal and professional growth in unexpected places is a ninja quality. While they're patiently delaying satisfaction, they're doing more than just waiting for a windfall. Ninjas watch and observe. They listen and learn to figure out what is really happening, and what they need to know and do to be smarter and more efficient.

Delayed Gratification in Public Policy

WHILE NINJAS ARE EXPERTS AT DELAYING GRATIFICATION AND using the intervening time wisely, society in general is horrible at it. We are an instant-gratification society, and I'll be the first to admit at least some of this is reinforced by the innovations made in the CE industry—all kinds of entertainment on demand, online shopping with free next-day shipping, etc. While this can be good for the economy, it's harmful in the realm of public policy. The consumers who demand instant gratification are also voters who demand instant gratification, so policy solutions tend to put a Band-Aid on our problems rather than fix them for the long term. Look at our own government for some examples of this. We refuse to address the serious problems facing entitlements, for example, because the best solutions are not immediate solutions.

Visiting India recently, I stayed in a magnificent hotel in the buzzing business center of New Delhi and journeyed by bus to Agra to appreciate the glory of the Taj Mahal. I was not disappointed by the world wonder, but the four-hour road trip each way left a deep impression. While the road itself was fine, the going was laborious. Animals, overcrowded jitneys, and even political demonstrations slowed and often stopped traffic. More, I kept waiting to get out of the "bad" area. Extreme poverty defined each kilometer without a break. Malnourished children were everywhere. Garbage lined the road. Plumbing and running water were rare, and the air was dirty from ubiquitous two-cycle engines. Electricity was sporadic at best. Life is tough in India, and survival is a daily challenge for many.

For India, the road to prosperity is long, and it is clearly based on education and innovation. The country has made a national commitment to taking the hard road, in the knowledge that investing time and resources in education will build a strong foundation for broader economic growth and prosperity.

As a result, India is now home to some of the world's most highly respected universities. There is an Indian joke that students who fail to get into the prestigious Indian Institute of Technology (IIT) settle and go to their "safety school"—the Massachusetts Institute of Technology (MIT). The truth is that Indians rely on education, as both an export and a strategy for growing their economy.

The United States also benefits from many highly trained Indians who emigrate to America for a better life. While an increasing number of Indian-Americans are born in America, many still emigrate here. In 2011, more than 69,000 Indian men and women obtained legal permanent resident status in the United States,[6] and the trend has been upward in the last decade—in conjunction with the technology boom.[7] Vivek Wadhwa, a technology entrepreneur and distinguished academic who studies entrepreneurial trends, estimates that more than 15 percent of Silicon Valley IT companies launched in recent years have been started by Indian immigrants.[8]

And now, homegrown Indian tech start-ups are beginning to threaten U.S. companies. Indian outsourcers are "performing sophisticated research and development," according to Wadhwa. "Today, Indian engineers design aircraft engines, automotive components and manufacturing plants, next-generation microprocessors, telecom products, and medical devices. Indian I.T. has grown from almost nothing in 1980 to an estimated $88 billion in revenue in 2011 according to Indian I.T. trade group, NASSCOM."[9]

You can visit any top-tier science class, read any American medical journal, or even watch the finals of the Scripps National Spelling Bee and see that America is benefitting from the Indian educational system, culture, and work ethic. Educated Indians help both India and the United States. Unfortunately for us, while many educated Indians emigrate to the United States, restrictive American immigration laws combined with a fast-growing economy and lower cost of living in India are making it less attractive for them to do so.

While India's growth has helped grow the world economy alongside the other three BRIC countries (Brazil, Russia, and China), it is taking some turns away from the free market that are impeding its success. For one, it is flirting with protectionism. In 2012, it toyed with the idea of restricting government telecommunications procurements to products made in India. While this may have boosted Indian manufacturing jobs and the local industry in the short term, it could restrict competition—forcing a higher price on lower-quality products. More, in the long term, it encourages other countries to impose similar requirements and raises barriers to trade. As India is a low-cost producer of products, an antitrade policy is not helpful for promoting exports or national economic growth. Fortunately, the government backed away from the idea.[10]

More, India has invested poorly in infrastructure. A basic role of government is to ensure clean water and air, and encourage investment in electricity and roads. India suffers from a lack of capital investment. India's size and growth make it attractive for investment, but its infrastructure obstacles discourage it. The temptation is to employ quick fixes, but if India can resist those temptations and continue to focus on patiently investing time and resources in real, solid growth, the final payoff will be so much better.

That said, India appreciates the value of knowledge and information, and it is reversing its historic legacy of being an impoverished country. It isn't any secret why Indians are sought-after workers for U.S. tech firms: It's because they have the requisite training and education (spurred by a national strategy) to do the work that is required. Meanwhile, America's K–12 education system falls further into decay.

The Patience Payoff

THE 1984 MOVIE *The Karate Kid* (YOU KNEW THIS WOULD COME UP, didn't you?) features a great example of how patience, perseverance, and delayed gratification prepare the ninja for battle. In the film, the protagonist, Daniel, wants to learn martial arts so he can protect himself from bullies. More, he wants to learn *right now*. But his teacher, the wise Mr. Miyagi, puts off Daniel's training. The famous "wax on, wax off" scenes show Daniel growing increasingly frustrated by the wait in his training while he is forced to perform seemingly irrelevant chores. But it all serves a purpose, as Daniel learns discipline, respect, and training techniques that serve him well later when those skills are put to the test.

The same payoff applies to modern ninja innovators who learn this important skill. Exercising patience is not only good for the project at hand, but it also trains and hones the information-gathering and analytical skills that come in handy when crisis hits. For ninjas, exercising discipline at all times meant less stress and less chance of mistakes in the execution of a dangerous mission. Using their "downtime" to train and gather information equipped them with a second-nature response to scan the environment and consider the possibilities.

Companies, industries, and governments should also scan the environment and be prepared for something that could devastate them. Seemingly minor decisions can have big consequences. Small problems can become big ones if not countered. Failure to respond quickly (and know *how to* respond) when you're attacked can be deadly.

Being fully prepared not only improves your work, but it also builds character and makes success worthwhile. In order to truly appreciate something, you have to earn it. The disciplines of patience, listening, information gathering, and education may seem unrelated at first glance, but in truth they work together to develop some of the key qualities of being a ninja, and a ninja innovator.

CHAPTER FIVE

THE ART OF WAR

The Never-Ending Battle

IN NOVEMBER 2010, ANDREW MASON, FOUNDER AND CEO OF Groupon, had an offer on the table for his company. As one of the hottest sites on the Internet since its founding in 2008, Groupon was at the top of its game. Its business model was simple: It offered consumers a daily deal at a local or national company. That was it. Mason was feted by the national media for such a ridiculously simple idea. He was on the cover of *Forbes* under the headline THE NEXT WEB PHENOM.[1]

It was precisely at this point, when everything seemed to be going so well, that Google offered Mason $6 billion for his company. As I'm sure you know, Mason turned it down. And that's when nothing seemed to go right. In June 2011, Groupon went public with a $12 billion valuation—double what Google had offered. But in February 2012, Groupon posted a $42.7 million loss for the previous year. By August 2012, the company's market value was just over $3 billion, which is half what Google had offered.[2]

Now, I'm not saying that Mason should have taken Google's offer. I'm also not saying he shouldn't have. Groupon is still around, albeit struggling, in a sea of competition, such as Google's own Google Offers—announced just months after Mason turned down the deal. I wish Groupon all the success in the world and hope it can weather these difficult times. But the Groupon example is a classic case of a business strategy that thrives as an early start-up but proves inadequate when it reaches for the next level.

Another way to look at it is that when it comes to developing and executing successful strategies, ninjas embrace the military adage that no strategy survives the very first contact with the enemy. Even the best strategies cannot fully anticipate what the competition will do, how rapidly they will do it, and how effective they will be. For your team to win after its very first contact with the competition, it must be talented and nimble: talented enough to recognize quickly what parts of your strategy are working well, working poorly, or not working at all, and nimble enough to adjust effectively. And the more thoroughly you have already thought through strategic contingencies, the more quickly you can adjust with a response that the competition doesn't expect or can't defeat.

And speaking of rapid, unexpected responses, I had an early lesson in that strategy that has served me well ever since. One day, while still of elementary school age, I was walking alone down a street and encountered the school bully. I don't recall quite how it happened, but seconds after he hit me, I punched him, giving him a black eye. From that day, he avoided me, which today I would define as effectively eliminating a competitor.

Psychologists say that bullies, like the cowardly lion in *The Wizard of Oz,* behave as they do in part to disguise a lack of confidence in who and what they are. Knowing this can be powerful, enabling you to respond in ways that your would-be tormentors never expect. One of my favorite personal examples occurred in the late 1980s, when Texas congressman Jack Brooks chaired the House Judiciary Committee. Chairman Brooks was known for being one of the most difficult members of Congress to deal with, and the fact that he had become chairman of such a powerful committee meant that he had won many more political battles than he had lost. With his bushy eyebrows and mustache, he reminded me of the outrageously funny and outrageously grumpy mid-twentieth-century comedy star "Groucho" Marx, whom most everyone recognized at the time.

At one point, Brooks was intent on quickly ramming through the House a so-called antitrust bill that would have needlessly hurt manufacturers and retailers across many industries. True to form, he refused to listen to our concerns, so I realized that I was facing another type of bully and would have to disarm him before he understood what was happening. So, I took a chance to be slightly outrageous myself by sending him and each of the twenty-one congressmen on his committee a "gift" that would certainly provoke a response—a then-famous plastic Groucho mask consisting of an outrageously prominent—and instantly recognizable—nose, eyeglasses, eyebrows, and mustache (they're still available today). Each mask was accompanied by a short letter that described why the legislation was ill conceived. My maneuver, in retrospect, a somewhat bold but definitely risky move, provoked an instant reaction from the committee members, with most of them doing the appropriate thing by reading Brooks's bill and considering our reasonable objections to it. In fact, I received many telephone calls from the members and their staff asking good questions about the dispute. And, as often happens, the bully's reaction when challenged was swift.

Less than a day after our Groucho Marx masks hit the congressional desks, Chairman Brooks called me into his office. Although our visit was not entirely pleasant, we quickly struck a deal: I agreed to support his legislation (with changes) in the House but was free to oppose it in the Senate. In return, we agreed that the other legislation that I supported would have to pass his committee. As a postscript, the deal I cut caused a bit of controversy among a few of CEA's corporate members, and their D.C.-based representatives tried to reverse the deal and get me fired. Fortunately for the legislation and me personally, such reps are rarely positioned to see the entire picture nor empowered to make decisions by themselves. Compared to some other trade associations, CEA works diligently to avoid putting lobbyists on our governing board, and our board has the final word about everything we do. Also, fortunately for me and CEA, my entire board of directors supported me in this high-stakes tussle. I was praised for cutting a deal that gave up to Chairman Brooks nothing but the "sleeves off my vest," as our board chairman, Joe Clayton (current president and CEO of DISH Network), described it at the time. Ninjalike, our legislative strategy was swift, bold, tightly targeted, and ultimately effective.

Process, Process, Process

OF COURSE, MOST BUSINESS STRATEGIES TAKE MUCH LONGER TO play out and regularly produce surprises that have to be dealt with quickly, before all the facts are in, thus increasing the risks of almost any reaction. I had another illustrative experience that involved the president of the United States in which I took a gamble that fortunately turned out well for him and me. It started when I received a call from a friend, Major General Mark Rosenker, who at the time was director of military affairs at the White House and was responsible for several thousand people who operated Camp David, Air Force One, and the critical, highly secure communications complex necessary for the president to stay connected with the world. Mark explained that then-president George W. Bush was leaving for Europe the next day on Air Force One and that the aircraft had a brand-new DVD player installed but no DVDs had yet been stocked. Evidently, he thought the head of the Consumer Electronics Association would have a warehouse full of DVDs at his fingertips. I assured Mark that I would immediately take care of the problem. We had a lending library through which our employees exchanged DVDs, and I knew that provided a basis for us to offer some movies to the president.

But I also went a step further. I wanted to provide the president with movies that he would actually enjoy. Unfortunately, the guidelines from the White House were a bit too broad: violence was okay, but no sex. So I called our librarian and asked her to research all information on the movie preferences of President Bush. She quickly provided suggestions, and then I went to Amazon and plugged in his favorite movies and received the Amazon recommendations for more current movies. I then took the list and drove to a nearby Best Buy and bought almost forty movies. Later that day I delivered to Mark a box of DVDs, carefully selected to meet the president's preferences.

If that were the end of the story, I would probably just save it for dinner parties. But for whatever reason—whether the White House staff was impressed with the quick response or whether President Bush really enjoyed the movies I selected, I'll never know—I received another call shortly afterward that would have decidedly more serious consequences.

The caller, Tom Campbell, Special Technology Advisor to the White House, explained that the Roosevelt Room across the hall from the Oval Office needed audio-video equipment and conference communications capabilities to be used in the event of any emergency, and asked if I could help supply state-of-the-art technologies. I quickly called senior executives at more than twenty CEA member companies who agreed to supply the necessary equipment as a donation to the National Park Service for the White House. We had to wait for President Bush to make a short visit to his Texas home before the workers could descend en masse to install everything, and they finished the job just minutes before he returned.

Only a few days later, on September 11, 2001, we were attacked by terrorists. The president had everything he needed to stay abreast of developments and communicate with almost anyone, anywhere in the world. Amazing timing? Perhaps a bit, but White House staff had the tools they needed in part because of CEA's version of a "quick-reaction strike force."

The reason I relate this story is that it illustrates one of the most important things that I've learned about strategic planning: Although your "final" strategy is important, it almost certainly will have to be revised again and again. For this reason, I think that your strategic planning *process* is even more important because, if done well, it helps prepare the team for the inevitable revisions. Because I knew my team and CEA members well from years of working together, I could rely on them to deliver great results in record time. Although I head the association itself, CEA's strategic power ultimately comes from the collective efforts of our members, especially members with CEOs who visibly lead and shape effective teams. Which brings me to another strategic success story that is amazing in very many ways.

Overcoming the Odds

NOEL LEE IS A FIRST-GENERATION AMERICAN WHO BEGAN HIS PROfessional career as a physicist working at the Lawrence Livermore National Laboratory in California. Before too long, he got tired of the grind, moved to Hawaii, and joined a rock band. That didn't go very well, but working with audio cables gave him an idea. So he returned to the mainland and decided to create a new company that would pioneer a new electronic product. He brashly named the company Monster Cable and gave himself the official title of head monster.

Before Monster arrived, people connected their component audio and video equipment with cheap, unbranded, generic cables. Noel believed that consumers did not recognize that their entertainment systems were only as good as their weakest link, the cheap cables, so he set out to upend the market with several innovations. First, relying on his physics background and high-quality raw materials, he created a line of cables that promised better, clearer sound and images. Second, he created a business model that enables retailers to earn much more by selling his cables compared to the generic ones. Third, Noel and his team (who are all available twenty-four hours a day) invest heavily in training retail salespeople on the benefits to consumers of using Monster cables.

The results have been spectacular—for the company, its customers, its team, and its retailers. In fact, it's not much of an exaggeration to say that Monster retailers love Noel, his brand, and his merchandising wizardry, which includes a live rock concert for them at every International CES show in Las Vegas. (Some dreams should never die.) Among Noel's triumphs is that he was the first to partner with the famous Dr. Dre to launch the large-headphone craze that has swept the world.

Also amazing is that Noel became disabled by a spinal condition that requires him to use a Segway to get around, but it doesn't stop him from traveling the world and navigating vastly different international cities, most of which are not as disabled-friendly as the United States. And he generously supports a foundation that provides Segways to our military veterans. I'm proud to say that when we first met, Noel was not pleased with something we at CEA had done, but together we worked it out. Now he is one of our best supporters and continues to be a font of innovative ideas.

I think the ninja lesson here is more a life strategy than any specific tactic Noel devised for Monster—although those tactics are clearly working. Noel went from a physicist, to a musician, to a businessman producing audio equipment. One can surely trace the progression here, but what's remarkable to me is that Noel never stopped revising his life strategy. Ever restless, always learning, Noel, a child of immigrants, transformed what would trip up lesser individuals—a personal disability and relative failure at a chosen profession—into motivators. When his strategy met the enemy and it didn't work, he changed his plans. It's no wonder Monster Cable is the success that it is today.

The Great Ones

WRITING ABOUT NOEL REMINDS ME THAT, UNLIKE HIM, MOST PEOPLE don't react well to change and try to resist it, and the same is true for businesses. But lives—and markets—are constantly evolving, so people and companies have to adapt or suffer the consequences. And, obviously, those who correctly anticipate transformations and prepare themselves are likely to be better off than those who have to scramble to adapt to changes already upon them. In consumer electronics, we repeatedly see the unfortunate differences. One company will successfully introduce a new technology and before too long other companies react by introducing similar products that offer little or no unique value to consumers and so are not successful, whether success is measured by market impact or returns to shareholders. I mentioned the tablet PC market in chapter 4 as an example of being prepared for battle. But what happened following the introduction of the iPad is also instructive for this chapter.

After Apple introduced its iPad in January 2010, more than fifty other companies introduced, or at least demonstrated, their own tablets. I saw them all at the 2011 International CES. In nearly every interview I gave about the show I fielded questions about the future of the tablet PC market. While I highlighted the explosion of iPad knockoffs as a great testament to the agility of the free market, I also predicted that only a few of those follow-on products would survive. History has verified that prediction. Of the dozens that were on hand, today consumers have a choice between perhaps five or six manufacturers. Why did so many of the others fail?

One reason is that most competitors are looking at the same markets, analyzing the same data, seeing the same trends, and coming up with the same answers. It sounds trite, but it is painfully true that it takes a special vision to spot opportunities, a rare courage to pursue them, and a uniquely skilled team to succeed. I vividly remember one firm (which will remain nameless) that was fortunate to have a CEO who, in his twenty years at the helm, increased the company's market value by forty times. In the ten years after he retired, the company's value essentially fell flat. Is it possible that one person can have that massive an impact on an already very large and diverse company? Yes, but only someone who has very rare talents.

That CEO was renowned for several trademarks, one of which was the quality of the firm's strategic planning. Unlike so many other executives, he knew that the only good reason for looking into the future was to do a better job today. His process forced timely decisions about which new ideas to pursue, which old ones should be abandoned, and which projects didn't need major changes yet. And the strategic thinking in his firm was a constant process, not merely a once-per-year exercise.

The late Peter Drucker, who may have been the world's greatest strategic management consultant, had one cardinal message for CEOs: "There is only one valid definition of business purpose: to create a customer. Because its purpose is to create a customer, the business enterprise has two—and only these two—basic functions: marketing and innovation. Marketing and innovation produce results; all the rest are 'costs.'"[3] In turn, creating a customer requires providing them with some value unavailable elsewhere. This is the core of competitive advantage and the essence of business strategy. I regularly think about this, and to me it is obvious that very few companies have found a way to create competitive advantages repeatedly. Take, as a current prime example, Amazon.

Brilliant on the Basics

OF THE iPAD COMPETITORS, THE ONE THAT IS DOING BEST APPEARS to be Amazon's Kindle Fire. I'm not surprised. As I mentioned in chapter 2, Jeff Bezos is one of those rare, wildly successful entrepreneurs who, in 1994, actually did start his company in a garage. But what more can be said about someone whose passion is taking a wrecking ball to staid industries, whose name generates four and a half million Google hits, and whose company, in fewer than twenty years, rocketed from zero to $48 billion in annual revenues? Well, one thing that I'm struck by is that Jeff seems to embody some of the same personal traits found in ninjas. To begin with, Jeff is anything but flashy, though he radiates a quiet intensity and has been called ruthless. Add to this his comment "You don't choose your passions, your passions choose you,"[4] and his very existence seems to strike mortal fear in the hearts of his competitors; he certainly seems to have the soul of a warrior. Not only is he famous for having told his shareholders that "it's all about the long term," but he also consistently wins most of his short-term battles.

And yet, I think he exemplifies the ninja for a much more fundamental reason—his passion for getting the basics right:

> We start with the customer and we work backward. We learn whatever skills we need to service the customer. We build whatever technology we need to service the customer. The second thing is, we are inventors, so you won't see us focusing on "me too" areas. . . . And then the third thing is, we're willing to be long-term-oriented, which I think is one of the rarest characteristics. If you

look at the corporate world, a genuine focus on the long term is not that common.

There are two ways that companies can extend what they're doing. One is they can take an inventory of their skills and competencies, and then they can say, "OK, with this set of skills and competencies, what else can we do?" And that's a very useful technique that all companies should use. But there's a second method, which takes a longer-term orientation . . . you ask, who are our customers? What do they need? And then you say we're going to give that to them regardless of whether we currently have the skills to do so, and we will learn those skills no matter how long it takes. Kindle is a great example of that.[5]

Jeff is one of those all-too-rare CEOs who focus on knowing the customers they have and the new ones they want to gain. In fact, according to Jeff, Amazon's explosive growth beyond books seems to have been triggered by its early customers: "We actually started to get e-mails from customers saying, 'Would you consider selling music, because I'd really like to buy music this way, and DVDs, and electronics?'"[6]

Jeff is regularly described as brilliant, not only because of his native intelligence, but also because he has led Amazon to so many technological breakthroughs and prospered in tough times like the dot-com collapse in 2000. I think part of his secret is that he's willing to challenge the status quo as well as shift his strategy when it's required. In its early days, Amazon was notorious for posting net losses each year. Somehow, Bezos convinced Wall Street to stick with him. He also had the temerity to take on the biggest players in his industry—Barnes & Noble and Borders—and win. (B&N adapted with a shift toward online; Borders didn't, and is extinct.) Now Amazon is knocking on the doors of other big names—Best Buy being the largest—because it wasn't content to just be a bookseller. With a few exceptions, Amazon has whatever you need.

It's a testament to Bezos's great ninja skills—adapting his strategy before each engagement—that he is one of the principal founders of the explosion in e-commerce, a medium that only gobbles up more of the retail market each and every year. Perhaps one day Amazon will face an enemy it cannot best, but so far that day has not come.

War of the States

THE GREAT NINETEENTH-CENTURY MILITARY STRATEGIST CARL von Clausewitz famously observed, "War is the continuation of Politik by other means." To close this chapter on the art of war, I want to discuss the impact of politics on ninja innovation. As we are painfully learning during our current, extended economic crisis, we need smart policies at the state level, not just on a national scale. I have some strong views about this because I've had a front-row seat watching policies work exceedingly well in the Commonwealth of Virginia, the home of CEA.

One of the virtues of our federalist system is that the states are left free (for the most part) to enact their own economic policies. What works in Vermont, for example, won't necessarily succeed in Oregon. But this system also has the added value of creating competition between the states for business. Some states are winning the fight, and some states are losing.

To most residents, Virginia is probably a great place to live because of its charming cultural mix of modernity and iconic history. But looking deeper, I think most would also agree that Virginians have been fortunate to have had a uniquely effective, bipartisan political system. That system has promoted an excellent business environment, which of course is also an attractive employment environment, especially for businesses on the cutting edge of technological innovation, like the members of CEA.

Indeed, I believe Virginia's "brand" has come to mean "best of class" in all the ways that matter most to our citizens. Through bipartisan common sense, low tax rates, high-tech government services, right-to-work laws, great universities, and a highly skilled population, Virginia has supported and attracted businesses to the point that our unemployment rate is now consistently well below the national average, and Virginia is currently ranked number nine in employment among the fifty states (as of May 2012).[7] Virginia has been named the best state for business by CNBC three times in the last five years, and leading American companies, including Northrop Grumman and Hilton Worldwide, have wisely relocated from the once-"golden" state of California to Virginia.

For decades, a bipartisan approach to fostering a business-friendly strategy has served the commonwealth well. Virginia has led the world in creating laws to promote commerce over the Internet. It also passed innovative laws encouraging funding of new Virginia businesses, and even used tax laws to support cloud-computing and data-hosting centers. Estimates show that half of the world's Web traffic flows through Virginia.[8]

Virginia's unique constitution has aided the creation and implementation of this strategy. Most significantly, the Virginia constitution limits and empowers the Virginia governor. Virginia is the only state that bars its governor from serving two consecutive terms. Although our governors may come to resent this restriction, it encourages them to lead decisively for four years, based on a long-term vision for the state's health, and without having their policies unduly shaped to get them through another election. One result is that we have had several all-star Republican and Democratic governors, many of whom moved on to a national stage. In fact, Virginia has more living former governors than any state—except perhaps Illinois if you count those currently serving time in prison—with nary a hint of scandal among them. Some of the more notable former governors include the nation's first elected black governor, Douglas Wilder; U.S. senators George Allen (R) and Mark Warner (D); and former national party heads Tim Kaine (D), Jim Gilmore (R), and 2009–2013 Governor Bob McDonnell (R). It can't be mere luck that they've all strengthened Virginia.

Shortly before I started writing this book, I was in San Antonio, Texas, to attend and speak at an important technology convention. One of the reporters asked me what San Antonio's leadership could do to enhance the city's position as a technology center. "Simple," I said:

> It can follow the example set by former Virginia Governor Jim Gilmore, who made the state's technology sector a priority of his administration from 1998 to 2002. Specifically, Governor Gilmore created the nation's first state Secretary of Technology, established a statewide technology commission, and enacted the first Internet regulatory policy that essentially said: hands off. Now, Virginia is known as a tech mecca, largely because of Governor Gilmore's foresight. San Antonio should

pursue a similar strategy that puts city government on the side of businesses and innovators. This doesn't mean central planning and picking winners and losers. Rather, it means focusing all policies toward the end of attracting investment and empowering innovators with the freedom and resources they need.[9]

Obviously, I believe this advice is what every state, city, town, and community should follow. Equally important, during Gilmore's 1998–2002 term, he enacted a diverse range of policies that strengthened the state. Among the most economically significant were a substantial reduction in automobile taxes, reduced spending by every state agency except education, education reforms that helped increase student scores on state and national tests, and increased funding for two of Virginia's primarily black universities. And, despite a national recession in 2001, Gilmore left behind $1 billion in the state's "rainy day" fund when he left office. Governor Gilmore has the further distinctions of being a graduate of the University of Virginia's undergraduate and law schools and a U.S. Army veteran (although not a ninja, he did serve as a counterintelligence agent during the Vietnam War).

I think it's clear that Virginia is a well-run state that prides itself on operational efficiencies that help keep tax rates low and on the ability of its governors in certain respects to reallocate budget funds without having to return them to the legislature. Virginia is also a textbook example of how a state can abide by the First Amendment and allow unlimited campaign contributions by requiring strict disclosure of all such contributions. We also have a training ground for potential political leaders at the Sorensen Institute at the University of Virginia (UVA). And speaking of UVA, Virginia has also had a bipartisan mission to build world-class institutions of higher learning. UVA, Virginia Polytechnic Institute and State University (Virginia Tech), George Mason University, and the College of William and Mary are all public schools climbing up the national leaderboard, especially with the ranking of California schools threatened by their state's out-of-control taxing and spending. (I'll discuss California in a later chapter.)

Virginia, starting with the Gilmore administration, stands as a great government example of how to pursue the art of war. Knowing it was in a battle for business with other states—and seeing so many states succumb to dumb antibusiness policies—Virginia saw an opening to exploit. It followed through with a smart, focused strategy to turn itself into a tech state. That tactic is working and I have no doubt that Virginia lawmakers will continue to refine their blueprint for whatever obstacles arise in the future.

Oh, one more thing: It's no coincidence that Virginia is one of two homes (the other being Coronado, California) to one of the best modern versions of ninjas, our U.S. Navy SEALs.

CHAPTER SIX

THE NINJA CODE

Honest Warrior, Happy Warrior

IN FEUDAL JAPAN, NINJAS WERE NOT WHAT WE WOULD CALL VIRTU-
ous warriors. When they first appear in the historical record they're
the personal agents of regional clans—specifically in the Iga and
Koga regions. Over time, as word spread of their skills and useful-
ness, the ninjas evolved into a professional class of mercenaries.
They founded their own guilds, kept their own hierarchies, and
otherwise competed with one another for the business of warlords.

By necessity, a mercenary cannot allow himself to be conflicted
about matters of right and wrong, good and evil. After all, it's bad
for business. If the evil warlord wants to hire you to help defeat the
enlightened ruler down the road, then off you go, happy to have
a job. You don't have the luxury of questioning your employer's
values or whether the war he has started is just.

Nor did ninjas concern themselves with "fighting fair," like their samurai brethren. Indeed, the ninjas based their entire profession on instilling in others the fear that they would not fight fairly. The conventions of warfare and rules of engagement, such as they existed in feudal Japan, were the very things that gave ninjas their mystique and power: If these standards did not exist, then the ninja lost his competitive advantage.

But even ninjas followed a code of conduct. In fact, their business depended on it. As highly specialized professionals, ninjas clearly had to be disciplined. Their skills were honed over years of intense study and training. Knowing what I know of martial arts, to be an expert in these skills requires a lifelong commitment. Only the most serious minds could expect to reach the professional level. It was to the benefit of the guild's leaders to demand among their subordinates a commitment to their craft and a standard of personal conduct. It made for better ninjas, which made for more business.

So the ninja occupied a unique moral territory: He was expected to break all rules of conventional warfare to achieve his mission but also was expected to follow a professional code of ethics. Naturally, today's ninja innovator should occupy the same ground—with some minor exceptions. Breaking the rules does not mean breaking the law, nor does it mean today's ninja should act like a mercenary, indifferent to matters of right and wrong. Above all, today's ninja should be a paragon of ethical behavior.

It's important to note that ethical conduct is not always the same thing as *legal* behavior. It is illegal to lie to the Securities and Exchange Commission; it is not illegal to lie to your spouse. Likewise, a CEO or entrepreneur who follows every law is not necessarily an ethical businessperson. The most famous recent examples are the unethical executives in the financial industry who (along with government officials and politicians) share the blame for the 2008 collapse of the housing market.

As many on the left continually point out, so far no one from Wall Street has gone to jail because of the financial crisis. The reason is simple: because no laws were broken. But that's not to say that the executives who played an active role in the housing collapse behaved ethically.

Whether it's the housing collapse or another example of business behaving badly, many identify greed as the driving sin that makes otherwise good people do unethical things. But I don't think greed suffices. Greed is just a motivator. Other motivators are providing for your family, achieving power, changing the world, and helping others. I find it helpful to know what motivates people, as everyone is different. I have no problem with those who want to make money, as securing wealth is not evil in itself.

Rather, it's when the pursuit of riches is accomplished through dishonest means that we have crossed into unethical territory. The ability to remain honest while succeeding is what makes a true ninja innovator. It might seem obvious, but naked honesty is rare, especially when we look at ourselves. This happens in both big and little ways. The little ways are the lies we tell ourselves about our height, weight, or increasing vulnerability as we age. We may ignore challenges in our work or professional relationships rather than deal with them. Honesty with oneself is a tough starting point for any ninja.

Self-honesty begins with emotional intelligence. That is a fancy term for looking at yourself from afar with candor and recognizing the motives behind your actions. If you can observe yourself and describe why you're reacting the way you are, you have come a long way toward real self-honesty. The opposite of self-honesty is self-delusion. We lie to ourselves to explain away our unethical behavior. And we're all pretty naïve when it comes to our own lies.

It's the same way with enterprises. Self-delusion is a commonplace flaw in any company or organization. This is especially true where the culture of an organization is such that obvious mistakes are allowed to go unchecked because no one is willing to speak up. Over a dozen years ago, I was visiting the Japanese headquarters of a major electronics company regarding issues with televisions and the V-chip. While I was there, the company's executive team asked for my thoughts on its huge investment in recordable CD hardware technology. I was struck by the unanimity among the team over their investment decision. They were all ecstatic about it. As any true ninja executive will tell you, complete, overexuberant unanimity on any decision, much less a major investment, is an obvious red flag.

When I expressed doubts about the consumer interest and market for their investment, they looked at me as if I was from another planet. (I was just from another country.) I don't think I'm ruining the surprise when I tell you that history proved me right. Recently, I was discussing this sad episode with one of the American executives of this company, who bought into the Japanese excitement. He acknowledged to me that the company had lost billions of dollars because it suffered from "groupthink"—a common malady in a culture that does not allow disagreement with the opinions of others, particularly of your superiors. *Groupthink* is just another term for self-delusion.

A more clinical term for it is *the Abilene paradox*. Coined by management expert Jerry B. Harvey in a 1974 article in the journal *Organizational Dynamics*, the Abilene paradox refers to a group of people who decide on a course of action that is counter to all of the participants' wishes. To illustrate the point, Harvey used an anecdote:

> On a hot afternoon visiting in Coleman, Texas, the family is comfortably playing dominoes on a porch, until the father-in-law suggests that they take a trip to Abilene [53 miles north] for dinner.
>
> The wife says, "Sounds like a great idea." The husband, despite having reservations because the drive is long and hot, thinks that his preferences must be out-of-step with the group and says, "Sounds good to me. I just hope your mother wants to go." The mother-in-law then says, "Of course I want to go. I haven't been to Abilene in a long time."
>
> The drive is hot, dusty, and long. When they arrive at the cafeteria, the food is as bad as the drive. They arrive back home four hours later, exhausted.
>
> One of them dishonestly says, "It was a great trip, wasn't it?" The mother-in-law says that she would rather have stayed home, but went along since the other three were so enthusiastic. The husband says, "I wasn't delighted to be doing what we were doing. I only went to satisfy the rest of you." The wife says, "I just went along to keep you happy. I would have had to be crazy to want to go out in the heat like that." The father-in-law then says that he only suggested it because he thought the others might be bored.
>
> The group sits back, perplexed that they together decided to take a trip that none of them wanted. They each

would have preferred to sit comfortably, but did not admit to it when they still had time to enjoy the afternoon.[1]

We've all experienced a situation like this in our personal and professional lives. When it's only a drive to Abilene, then it's harmless enough. But a company that takes the proverbial drive to Abilene could severely, if not fatally, harm itself. To avoid the Abilene paradox, enterprises must not only have employees willing to speak out against a bad idea but must also have an executive team willing to listen.

I will relate a story that isn't very flattering to one of the greatest ninja innovators in our time, but it's instructive nonetheless. At one International CES some twenty years ago, Microsoft's Bill Gates was the keynote speaker. During Gates's rehearsal, I was the only non-Microsoft person in the room. Listening to Gates practice, I had a number of suggestions that I thought would improve the presentation. Assuming my opinion would be valued, I shared my opinions with a couple of Microsoft people. After hearing what I had to say, they looked at me nervously and suggested that I speak directly to Gates. No problem.

I entered the room where Gates was with a few of his key people and said I had some suggested improvements on his presentation. As a successful CEO, he immediately recognized the value of neutral input, and he listened to my suggestions, accepted a few, and discounted the rest. That was fine. I should have left it at that.

Instead, I went beyond where I should have ended the conversation and blurted out that I thought the product he was introducing at the show would be one of Microsoft's most significant products ever. I said I was very impressed with it and thrilled he was choosing to launch it at the International CES.

What I didn't know until the next day was that the product was the baby of a special employee at Microsoft, Gates's then-girlfriend Melinda, who would later become his wife. Upon reflection, I surmised it was a delicate internal situation for the top people at Microsoft, who were likely reluctant to give Gates any real feedback or input on the product or how it was being launched.

The product was introduced as Bob but became known—derisively—as the Dancing Paper Clip (DPC). See? I bet you know immediately what I'm talking about. Aimed at teaching users how to use Windows products and capabilities, many users found the DPC an annoying distraction. Even worse, it was difficult to get rid of and was always popping up at the most annoying moments. It soon became the butt of late-night TV jokes and didn't reappear ever again on a Microsoft product.

The lesson for me was an embarrassing one. First, while my input on how to give an effective presentation was valuable, even to a master of presentations like Bill Gates, my input on a software product in this case was totally off the mark. Second, as successful as you may be, you need to be careful if your top leadership believes they can't have an honest discussion with you. In this case, the founder's relationship with an employee may have made criticism of the product off-limits. Third, even a successful company like Microsoft makes mistakes—embarrassing failures. The important thing is to learn from them and keep taking risks.

We are also experiencing a collective Abilene paradox as a nation. As I write this, the United States government is in denial about the scope of its financial problems. The Democrats and Republicans cannot even agree on the facts defining the financial problem, even though the facts are clear.

No Labels (nolabels.org), a group I have been affiliated with for years, has proposed several commonsense solutions to encourage communication and honesty in and from government. Among them is a specific suggestion that the president and congressional leaders from both parties sign a statement of financial facts each year so they can agree to have a policy discussion from the same baseline. It used to be that journalists and the media would be neutral arbiters of the truth. Sadly, that is no longer the case. Like our politics, our news is hopelessly divided along partisan lines, with each side preferring the networks and outlets that cater to its version of the facts.

To be effective as a ninja leader one must establish a culture of honesty within one's team. A leader cannot tell his team that the mission before them will be easy if in reality many of them won't likely return. A leader's honesty is almost always rewarded, even if bad news has to be delivered, with increased loyalty and improved performance. Likewise, if a leader is honest with his team, the team is more likely to be honest with its leader.

At CEA, I try to reward honesty by conveying an organization-wide message that honesty will never be punished. I want my team to tell me the hard truths. Recently, after a speech, I asked a new employee for feedback on how I could improve my presentation. She responded that I did well but had appeared uncomfortable and awkward in a certain section of my presentation. She was right. Her star rose in my eyes. I thanked her profusely because I needed candid feedback. It's easy to tell someone that they are wonderful; it's more difficult and ultimately more important to tell them how they can be better.

Honesty is a ninja virtue. But like many virtues, it must be practiced in both personal and professional settings. When an individual, enterprise, or country is honest with itself—when it confronts the truth instead of choosing to believe comfortable falsehoods—then it sets itself on the road to lasting success.

Winning Isn't the Only Thing

A FAMOUS SPORTS MAXIM STATES, "WINNING ISN'T EVERYTHING. It's the only thing." No one is quite sure who first said this, although Vince Lombardi was known to repeat the phrase often. In athletics, the maxim is universal; there are only winners and losers in each game. It makes sense to apply this to success in other ventures as well. But only to a point.

No ninja innovator is going to win every battle. In fact, as I've indicated elsewhere, you're probably going to lose more often than you win. There are enterprise-wide losses, and there are personal losses: the job you didn't get, the job you lost, or the client you couldn't lock down. In these instances, winning is certainly not the only thing. In many ways, it's how you act or perform after the contest is over that is far more important.

For me, when colleagues are down, I do my best to reach out to them. If they lost their job, I call. If someone has a bad story written about them in the press, I ask how I can help. I try to do what I can to help someone when they are truly down and others are abandoning them. People don't need the help when they're up; it is most appreciated when they're down. A true ninja innovator is always thinking long term. You can make a genuine human connection with someone when they are down—despite personal differences on business or policy issues—and that connection will go a long way toward fostering a more effective relationship with that person in the future.

Choosing whether to win or lose can also be an effective strategy. Sometimes letting someone else win in the short term simply makes sense. One common strategy is to let your competitor go first to market. You let them make the investment and learn from their mistakes. You let other companies introduce new products first to the market. You then watch carefully and learn from your competitors' mistakes and only later introduce a competing—and usually better and less expensive—product.

Another time when ninjas choose not to win is when they are focused on the long-term victory rather than the short-term win. It may make sense to let your competitor win to build up their confidence; indeed, an overconfident competitor is a weak competitor. Having a competitor underestimate your strength can be a competitive advantage.

Take one of the most phenomenal success stories of the past decade: the rise of Facebook. The site is so ubiquitous these days that we tend to forget that Facebook wasn't at all unique when it first appeared as a social-networking platform for college kids. Other sites like Myspace and Friendster had already tilled the ground that Facebook would go on to dominate. Indeed, in Facebook's early days, Myspace and Friendster took pleasure in *ignoring* Facebook, preferring to define the market as a competition between the two of them.

Which was just fine for Facebook. Had Mark Zuckerberg publicly declared war on Myspace and Friendster in Facebook's early days, then those two behemoths might have taken notice and squashed the upstart (or simply stolen its ideas, which were proving better and more in line with the evolution of social media). As it was, neither Myspace nor Friendster considered Facebook a true threat. This allowed Zuckerberg to monitor carefully both sites to see what they were doing right and what areas they could improve upon. The best part? Myspace and Friendster felt no need to change their approach. They weren't threatened by Facebook, so they didn't think anything was wrong with their services.

But, boy, were things so very wrong. Friendster, the original Facebook, was slow and unwieldy. Myspace made the mistake of throwing everything but the kitchen sink at users' profile pages, which resulted in an assault on the senses for the uninitiated. Moreover, Myspace soon became a site catering to adolescents, which not only alienated the older crowd but also fueled the reputation of the site as a target location for sexual deviants and pedophiles. And while these two Goliaths fought it out, Facebook chugged along, stealing the best parts of both sites and innovating others, while avoiding the missteps that no one really considered until Facebook exploded in popularity.

So while Facebook won in the end, it knew that a quick victory wasn't the most important goal in the beginning. Getting it right was far more crucial.

And then there are the times when the wiser course is not to take the field at all. Often it is best to avoid a real or contrived match against a competitor. It implies and demands that there will be a winner and a loser, when winning and losing won't benefit either company. A classic in this case was the very public, very heated feud between Steve Jobs and Adobe over the latter's Flash technology. We don't need to go into the reasons why Jobs hated Flash, but loathe it he did. He told his biographer Walter Isaacson, "Flash is a spaghetti-ball piece of technology that has lousy performance and really bad security problems."[2] Now, I have no idea what a "spaghetti-ball piece of technology" is, but I'm pretty sure I wouldn't want Steve Jobs calling my product that. But as the feud played out very publicly in the media, it began to look less like a critique of a certain technology and more like a personal vendetta for Jobs.

Consumers still gobbled up Apple's mobile devices, the iPhone and iPad, despite their lack of Flash capability. And yet, the feud allowed Apple competitors, such as Google's Android software, to advertise their products' Flash capabilities. The fact is that Flash is the standard in the industry. If you don't have it, then your browsing experience is greatly compromised. In the end, it's hard to see how the Apple-Adobe fight benefited either company, and it stands as an example that sometimes it's best not to fight at all.

As a Washington trade organization, CEA is often battling other industries. Yet in every case, I know and work well with my opponents. I reach out to them when our issues coincide, and I battle them fiercely when we disagree. I invite them to our events and join in recognition of them publicly when appropriate. And sometimes when their CEOs lose their jobs, even then, consistent with my ninja approach to life, I reach out to them as individuals and share my support in their time of need.

Of course I realize that a trade association is a different beast than a company. Our job is to promote the interests of all of our members while remaining agnostic on particular platforms or products. Even when CEA engages an opponent whose interests contrast sharply with our members', my goal is not to destroy the opponent. Some of my members might disagree, but I'd rather see the U.S. economy full of healthy, vibrant, and competing industries than full of dying industries.

Like the ninjas of feudal Japan, today's ninjas do not and should not live in a black and white world where everyone is either a good guy or a bad guy. The truth is that everyone views themselves as a good guy, as the guy just trying to get ahead or stay afloat. I might protest loudly against an industry that uses artificial means to stay alive—such as the steel industry or the broadcasting industry (which I'll get to in a later chapter)—but I do not view them as the enemy. Ninja maturity is the ability to see all shades of gray, to deal with nuance, and to understand that sometimes we must live with ambiguity.

Someone once said, "Show me someone who doesn't mind losing, and I'll show you a loser." The ninja's approach is different. The ninja can accept losing, because losing itself can and should be a learning experience.

But, of course, winning is the goal; there is nothing sweeter than a hard-fought battle where, when the smoke clears, you have a clear win. It is even sweeter when the victory comes because of creativity, confidence, and hard work rather than through a weak competitor, government intervention, or good luck.

For a ninja, winning is not everything. Victory happens when you believe it is essential. But even before you conclude that you must win, you should consider if and how you can work with your competition. I have found that "co-opetition" is much better than competition. The old saw that you should "keep your friends close and your enemies closer" is true for ninjas as well.

Not All Rules Are Equal

THE VERY BEST NINJAS FOLLOW A CODE OF CONDUCT—BUT ALSO break the conventional rules of the game. They discover ways to surprise their competition. And it's very common for an out-smarted competitor to accuse the ninja company of unethical or illegal practices. At this point, government usually enters the picture. It's an unfortunate reality in our present circumstances that government has a hand in everything business does. Barely a news cycle passes without the government investigating a successful company for real or imagined transgressions.

For years, Google famously didn't have a Washington lobbying office. But then Google became wildly successful and things changed. Competitors began to envy Google's success, and they called on their most trusted allies: their friends in Congress or the White House. When the government started going after Google—no doubt spurred by well-heeled competitors—Google succumbed to reality. In 2005, Google announced the opening of its Washington office on its blog: "It seems that policymaking and regulatory activity in Washington, D.C., affect Google and our users more every day. It's important to be involved—to participate in the policy process and contribute to the debates that inform it. So we've opened up a shop there."[3]

Which is fancy PR-speak for: Fine, we'll play it your way. Today, Google has an extensive lobbying arm and in the first quarter of 2012, the company spent a record $5 million on lobbying Congress—which was more than Microsoft, Apple, and Facebook combined.[4]

This is the usual route for any large, successful company. Its initial success is marked by a period of government naïveté, which is quickly followed by a hard lesson in the ways of the world: You want to play, you have to pay. These days everyone is going after Google through private and public channels. Google's only recourse is to fight fire with fire.

But there's something more sinister in the need for companies to bend the knee to Washington politicians. The expansion of government into all facets of the free market has led to perverse antibusiness practices. There's very little restriction these days on what government can do to a private enterprise. A couple recent examples will suffice.

On August 24, 2011, federal agents armed with automatic weapons stormed several Tennessee factories and the corporate headquarters of Gibson Guitar Corporation. It was the second time in as many years that Gibson had been raided by the government. The reason? Federal agents from the Fish and Wildlife Service believed Gibson was illegally importing protected wood from far-off places like India and Madagascar. Never mind that both countries insist that Gibson did nothing wrong. The famed guitar manufacturer has not been charged with a crime. But this still didn't keep the federal agents from walking away with $500,000 in Gibson products.

Behind the raids is a law known as the Lacey Act, which regulates trade of endangered plants and animals.[5] First passed in 1900 to regulate only wildlife, the Lacey Act was amended in 1998 to include trading in wood that has been illegally logged. While the 1998 amendments were aimed at disrupting illegal logging, they also broadened the law to criminalize anyone unknowingly possessing certain rare wood or plant materials.

And that's the major problem with the Lacey Act. It is broadly written with too few process protections and places the burden of proof on the owner, not the government. The law puts every American who buys antique and wooden products at risk, from musicians to manufacturers to dealers in antique furniture. Some musicians now refuse to travel in or out of the country with vintage instruments, which could be subject to seizure.

Gibson CEO Henry Juszkiewicz explained his frustration: "[The Lacey Act] has nothing to do with conservation, nothing to do with how scarce or not scarce rosewood or ebony is. It has to do with jobs. They have raided us, come in with weapons, and they seized $500,000 worth of product. They shut our factory down and they have not charged us with anything at this point."[6]

Henry Juszkiewicz is a true American success story. Born in Argentina, Juszkiewicz rose from the bottom, got his M.B.A., and bought a struggling Gibson in 1986. He successfully turned the iconic American company around and expanded its sales. He went on a buying spree and bought the jukebox company Wurlitzer and the iconic piano company Baldwin, and he recently bought the Japanese receiver company Onkyo.

Like every other ninja innovator, Juszkiewicz makes mistakes. He had an idea for connecting all electronic devices in a handshake system requiring product registration. It would have satisfied record company piracy concerns but was complex and not embraced by electronics companies. He tried and failed but went on to make his company bigger and better.

Of course Juszkiewicz's biggest challenge today is not his business—it's the federal government. Instead of meekly bowing to the federal bully, Juszkiewicz innovated in his own way by responding differently. He stood up to the bully. When the federal government refused to explain or apologize for its actions, Juszkiewicz went to the media. He told his story in print, on the radio, on television, and on blogs. Speaker of the House John Boehner invited him to attend a speech President Obama gave before a joint session of Congress in September 2011—a clear message to the Obama administration that business owners had had enough.[7]

Juszkiewicz says it well: "It's not easy to compete on a global basis. We're competing with Chinese, European companies. We'd like to feel like we're respected and help create jobs."

Exactly. What would possess a federal agency to storm a successful, iconic American company on the flimsy excuse that it *might* be illegally importing protected wood? Does that require federal agents armed with automatic weapons to storm private property and cart off hundreds of thousands of dollars of product? This has to change.

Rules are necessary, but if they are ambiguous, unclear, or give too much discretion to their enforcers, they stifle business and innovation. Juszkiewicz's case proves he has been wronged. He has put too many resources into defending himself from the government and paying legal fees and a fine, when he could have spent those resources on growing his business.

While Juszkiewicz and Gibson settled the case with the U.S. government in August 2012 for a few hundred thousand dollars, it is clear the settlement had little to do with an intentional legal violation and everything to do with Juszkiewicz trying to save his business from fighting the federal government in the courts. The Gibson press release quoting Juszkiewicz says it all:

> We feel that Gibson was inappropriately targeted, and [that this was] a matter that could have been addressed with a simple contact by a caring human being representing the Government. Instead, the Government used violent and hostile means with the full force of the U.S. Government and several armed law enforcement agencies costing the taxpayer millions of dollars and putting a job-creating U.S. manufacturer at risk and at a competitive

disadvantage. This shows the increasing trend on the part of the Government to criminalize rules and regulations and treat U.S. businesses in the same way drug dealers are treated. This is wrong and it is unfair.[8]

And then there's the famous example of Boeing, another iconic American company. In March 2011, the National Labor Relations Board (NLRB) filed a complaint against Boeing for opening a plant in South Carolina instead of Washington State. In effect, the NLRB was trying to stop a private company from doing business. Why? Because, said the NLRB, Boeing was just "retaliating" against Washington State union workers, who had gone on strike in the past. Since South Carolina is a right-to-work state where workers can *choose* whether to be in a union, Boeing allegedly was just trying to stick it to Washington unionists.

It's a dangerous precedent when the federal government says where a company can and can't open its factories. The South Carolina plant was going to generate thousands of jobs, but to the Big Labor allies at the NLRB that was less important than staying in good standing with one of the Democratic Party's largest donors.

After the issue was finally resolved and the complaint withdrawn, GOP presidential candidate Jon Huntsman summed up the sad, sorry episode best when he said, "The NLRB decision is a victory in a battle that should have never been fought. Their action against Boeing in South Carolina was an unprecedented attempt to interfere in the free market, and an attempt to politicize companies' decisions as [to] how and where they create jobs."[9]

Even if they break the rules, ninjas follow the law. That doesn't mean that the laws are just or sensible. If the laws are unclear, arbitrarily enforced, harmful, or simply unethical, ninjas will find a way to put their case to the people, just as Juszkiewicz and Boeing have done. But we shouldn't forget what their experiences suggest about our supposedly free-market system. A government that actively works against the interests of business and job creation is a government that is immoral and unethical—and the enemy of ninja innovators everywhere.

Ninjas Pay It Forward

AS A FINAL NOTE ON THE IMPORTANCE OF BEHAVING ETHICALLY, WE cannot overlook the value of sharing wisdom with others. Mentoring, formally or informally, is both an obligation and a joy for anyone who has succeeded.

The most successful organizations have a culture of sharing and training. This means that the more experiences and success a person has, the greater the obligation to impart wisdom and pay it forward to the next generation. Sharing with, mentoring, developing, and teaching others are what a ninja must do to achieve the marks of a great leader and become a successful person.

Ninjas should also know that the very act of teaching and engaging with others is itself a broadening experience. Although it may seem like mentoring and sharing are unselfish acts, in reality they are wonderfully joyous and produce great satisfaction for the mentor.

A corollary to the value of mentoring is that you're never too old or too successful to be taught yourself. I recently had lunch with an accomplished and slightly older CEO who had served as board chair of the corporation at which I had my first senior position. Not having seen him in years, I was delighted to catch up. But I was exhilarated when he expressed personal interest in my career, observed my success, and challenged me to reach higher. Upon parting, I blurted out awkwardly, "Thank you so much for your advice. It was so great to hear, as all my mentors have died."

I wasn't just trying to be nice; what I said was entirely true. Other than my wife and a few close friends, I am—and have been for a few years—mentorless. It's an unsettling experience, particularly because I have been blessed with so many wonderful role models in my life. If anything, this realization has pushed me to continue sharing my wisdom and mistakes with others, who someday will be able to pass it on to an even younger generation.

Several years ago I enjoyed a memorable dinner in Las Vegas with the two people whose advice I valued most: my father, Jerry Shapiro, and my mentor and friend Jerry Kalov. I loved both men and will never forget that evening sharing stories and discussing life. Jerry Kalov, while running a public company himself and serving as my lead board member, told me that despite early successes, I would fail, and he would stand by me when I did. This allowed me to take risks, and of course many of these risks led to spectacular failures. He supported me, and I grew from each failure.

Sadly, within several months of that dinner, both Jerrys passed. I often wonder what questions I would have asked had I known their time was so limited, what they would have said, and what advice they would have given. Even today, facing tough situations and decisions, I often ask myself: What would Jerry say?

As I counsel and mentor younger CEOs, I encourage them to never stop seeking guidance from others, even those not in their industry. After all, you never know what you can learn from a trusted outsider—someone whose only stake in your success or failure is based on pure friendship. As I said in chapter 4, ninjas should never stop listening to others.

Above all, this means staying engaged. Continue making friends, look for others in your organization to mentor, and never forget that you are where you are because someone helped you along the way. Being mentored and mentoring is part of the life process of giving back. Even I, a lifelong learner who knows that luck plays a role in life, realize the amount I don't know increases with my maturity.

Every tae kwon do black belt on their way to becoming a true ninja will also have to be a teacher to the less experienced students. Teaching classes, even at a lesser belt level, I was so conscious that my form had to be perfect, my kicks sharp, my katas precise. You learn so much more how to do something well when you teach it. Which is another way to say that when you mentor someone else, you're also helping yourself.

CHAPTER SEVEN

NINJAS BREAK THE RULES

Unlike their feudal counterparts like the flashy samurai, ninjas were not constrained by Bushido—the "way of the warrior." Similar to the chivalric code that governed the conduct of European knights, Bushido defined the terms of both an honorable life and an honorable death—two concepts not particularly valued by the ninja.

In our modern parlance, we would say ninjas were significantly more "results oriented" in their approach to tactical challenges. In fact, while we think of the two classes of fighters as being separate and at odds with each other, at times they had to work together. The samurai, honor-bound by Bushido, could not accomplish certain tasks. Instead, they had the ninjas do it for them.

Bushido, like most chivalric codes, was a considerably good thing. We appreciate when people carrying swords and riding around on horses feel some constraints regarding keeping the peace and protecting the weak (even if the code allowed the samurai to cut down any commoners who insulted them).

On the other hand, a rigid adherence to traditional methods is a huge hindrance to innovative problem solving. Once "the way it's always been done" stops working, a ninja innovator doesn't hesitate to try new things.

In the case of premodern Japan, it was precisely this lack of openness to new ideas that dropped the insular island so far behind the comparatively open Western world by the 1800s. Perhaps if the ruling class had welcomed a few more ninjas, Commodore Perry's gunships wouldn't have wreaked such havoc upon their arrival.

California: Not a Ninja State

SPEAKING OF PLACES THAT NEED A GOOD WAKE-UP CALL, LET'S look at California.

California is home to hundreds of the world's most innovative technology companies. More than 10 percent of the Fortune 500 calls California home, with heavyweights like Apple, Cisco, Google, HP, and Intel topping the list. Facebook started at Harvard, but you'll find its corporate headquarters in Menlo Park, California. Even large technology companies that aren't based in California end up having a significant presence there. Washington-based Microsoft operates a Silicon Valley campus that is the company's second-largest facility. More than two thousand Microsoft employees work in the state.

California also remains the start-up capital of the world. Start-ups in California routinely pull in 40 to 50 percent of all venture capital funds invested in the United States.[1] Entrepreneurs know where the money and talent are concentrated, so many of the best and brightest flock to Silicon Valley or the Southern California megalopolis to develop and launch their ideas. This in turn draws in more investors in a self-perpetuating and virtuous cycle.

So why in 2011 did the state rank fiftieth in the nation in the category of new business creation?[2] And how can a state with such a dynamic business community sustain an unemployment rate over 10 percent, among the highest in the nation?

For starters, in many ways California is resting on its laurels. Silicon Valley became what it is today because of its proximity to Stanford University. Around the middle of the last century, Stanford administrators pursued a strategy of sending out the school's exceedingly bright students into the surrounding area to build their tech companies—mostly as a way to counter the East's dominance in business and industry. The strategy worked, which meant that if one wanted to start or work at a tech company, Silicon Valley (as opposed to New York or Chicago) was the place to go. Like a snowball rolling down the hill, this trend became an avalanche by the time computers had reached the stage of consumer interest— the 1970s.

Even today, if one wants to hobnob with the smartest techies in the world, one goes to Silicon Valley. That's where the technical, financial, and structural advantages continue to be. But for how long? The state's politicians quickly began to view California's success as a birthright—something inherent in the state's DNA—and so they went to work sucking the state dry. But there is no law of nature—well, except for the weather perhaps—that says California *must* be the center of the tech solar system. Indeed, if recent trends persist, it might, in a few short years, be as distant as Pluto.

For years California's state government never met a tax it didn't love and a regulation it wouldn't approve. For almost a decade, *Chief Executive* magazine has ranked California as the worst state in which to conduct business, pointing to excessive government regulation of businesses as one of the key reasons the state fared so miserably. This isn't an outlier, as California ends up on the bottom of the rankings for virtually any group that evaluates states on their business-friendliness.

Whole companies meanwhile are fleeing the state for more friendly climates. Northrop Grumman left in 2011, meaning Southern California is now no longer home to even a single major military contractor (many of the jobs did stay behind). Nissan North America relocated to Nashville, Tennessee. As for the companies that haven't left yet, they have already made decisions about investing and hiring in other states, leaving California out in the cold. According to one measure, 254 major California companies shifted significant numbers of jobs or investments out of the state in 2011.[3]

These departures and realignments are showing up in the employment numbers. From January 2008 through January 2012, the state lost more than 850,000 private-sector jobs, easily the worst showing of any state.[4] And it's not just because the state is so big; only seven states had a higher percentage of jobs lost.

California's tax burden is high, but the effects of overregulation can be even more crippling to business success. The California state legislature commissioned a study in 2009 that found government regulation costs the state a staggering $500 billion per year—almost one-fourth of California's entire gross state product.

The necessity of dealing with all these regulations has spawned a cottage industry of lawyers and consultants who help large corporations stay on the right side of the law. For California employers it's an added cost of doing business. Unfortunately, for small businesses with lower profit margins, these additional costs can be the difference between making payroll and going out of business. Compliance and opportunity costs inflicted by complex regulations cost small-business owners thousands of dollars each year—money that could be invested back into their businesses and the local economy.

The consumer electronics industry has battled out-of-control California state environmental regulators for years, even though we share their goals and have spent billions of research and development dollars on products that can perform better with fewer environmental impacts.

I'll give you an example. California law mandates a balance between the benefits and burdens of new regulations for consumer products by requiring that new rules "not result in any added total costs to the consumer over the designed life of the appliances." However, for the last six years, the California Energy Commission (CEC) has evaded these requirements by simply gaming their analyses and relying on obsolete data—in an industry where change happens literally overnight—and by using unrealistic assumptions. This systematic bias allows the CEC to claim phantom energy savings will reduce operating costs, netting out higher product prices on everything from battery chargers to televisions.

As mentioned, the oddest thing about California's incessant thirst for new regulations is that so much of what the state government does is totally unnecessary. The consumer electronics industry is already at the forefront of energy efficiency and sustainability, because our customers want batteries that last longer and products that cost less to operate.

Among the industry's existing sustainability programs are eCycling initiatives, green product standards, and efforts to educate the public on energy efficiency trends and opportunities. These innovative, proven approaches have resulted in significant energy savings over the years. As a result of innovation, competition, and the federal government's Energy Star program, the amount of electricity needed to power an LCD television set fell 63 percent from 2003 to 2010. Unlike the CEC's energy-usage mandates, Energy Star is an innovation-friendly program that encourages both competition and consumer choice.

The CEC recently gave itself permission to pursue new regulations for a wide range of high-tech consumer products and IT equipment, including computers, displays, game consoles, imaging equipment, servers, and set-top boxes. It doesn't matter to the CEC that successful energy-efficiency programs are already in place for all of these product categories; its motivation is purely increased control and the result is unnecessarily higher costs for manufacturers, retailers, and distributors.

There is some hope on the horizon, as the California legislature considered a bill in 2012 that would have curbed these sorts of abuses, specifying, among other requirements, that the CEC would have to rely on the most current data available for all proposed regulations and would have more flexibility to eliminate unnecessary and outdated regulations.[5] This being California, environmental and electric utility lobbyists killed the legislation as it was about to pass at the end of the 2012 session. It is incredible that environmental groups and utilities opposed legislation that would lead to more meaningful and rigorous rule-making yielding real, not phony, energy savings, but this is sadly true. The Natural Resources Defense Council (NRDC), for example, simply does not care about science or good analysis as long as it can push new restrictions on technology. Moreover, a few California legislators reflexively do whatever the NRDC wants, and thus California continues to wear the crown of idiocy as the worst state in which to do business—science be damned.

It is not only technology manufacturers who feel the brunt of California's overzealous regulation. California is also home to many of the nation's top video game companies. Yet the state passed a law banning the sale of certain video games to minors, despite the fact that video games now carry parental warning labels. Thankfully the Supreme Court threw out the law on First Amendment grounds, ruling that a state could no more ban the sale of a video game than it could the sale of a book or other piece of artistic expression. (In a rare example of justice and common sense, California also was required to reimburse nearly a million dollars in legal fees to the winners.)

As California does everything it can to hamstring the private sector, it lavishes money on the public sector. The state is home to a huge population of unionized government employees who believe large defined-benefit packages, ever-higher salaries, and restrictive work rules are their birthright—even as their neighbors in the private sector are getting squeezed on all sides. California's unfunded liabilities for state and local pension systems are at a choking $500 billion. Between 1999 and 2012, pension costs grew 11.4 percent a year.[6]

The scariest aspect of California's state government is that its only accountability seems to be to itself. For years, the union representing prison guards has pushed for legislation that would increase prison populations. California had some 150,000 inmates in state prisons, although that number has dropped since the state began pushing low-level offenders into local jails to comply with a U.S. Supreme Court mandate to reduce overcrowding.[7] (By comparison, Greece, which has a similar population and significant social challenges of its own, has only about 12,000 people locked up.) Prisons still account for about 9 percent of California's budget.

California violates ninja rules. Even in the face of clear and present danger to its budget, bond rating, and corporate climate, the state's government has refused to adapt to change. The state legislature continues to favor the minority of union workers and trial lawyers who together impose absurd costs on business and on taxpayers.

Once a primary driver of U.S. prosperity, with its technological wizardry, booming ports, and heavy industry, California is quickly becoming the weight that brings down the rest of the country. Somehow, with only 12 percent of the U.S. population, California manages to be home to one-third of all Americans receiving welfare. More, in 2012, it owed the federal government over $9 billion for unemployment payments—more than another 47 states owed collectively!

To date, California has stayed afloat through a combination of federal bailout funds, budget gimmicks, and other quick fixes, hoping that in the long run things will get back to the old normal. They won't. The world has changed, and California needs to change with it if it's going to again become the type of place that rewards success and welcomes great businesses and citizens who value their liberty.

California's saving graces are its phenomenal beauty, lengthy coastline, and delightful weather. People will always want to live there. But like a person gifted with tremendous beauty, California cannot subsist—much less flourish—on its physical appeal alone. It needs to cut programs, draw down the government workforce, and reduce taxes and rules that smother business. Restricting spurious lawsuits and eliminating programs redundant to federal efforts would lift even more of the burden off taxpayers and employers. The state could take an immediate first step toward fiscal sanity by cutting its enormous prison population through a combination of pardons and legal changes for victimless crimes.

As a final note, the smartest leader in the world cannot manage California to fiscal health without the support of a legislature willing to make fundamental change. Until and unless that happens, California will continue its long descent from past glory.

There's No Such Thing as Too Big to Succeed

ONE OF THE EXCUSES PEOPLE USE TO EXPLAIN AWAY THE PROBLEMS of governance in California is the state's size. The same excuse gets made for companies. This is bunk.

Large corporations do face unique challenges, starting with the fact that they are, well, big. A great product idea that might be developed and released for testing in a week at a small Web start-up might never get past the first round of approvals at a large technology company. Middle managers often avoid advocating for anything that might be risky and could jeopardize their careers. Good ideas routinely get squelched on their way to top leadership.

Big companies also frequently exhibit the slavish adherence to tradition that prevented the modernization of Japan up to the nineteenth century. Companies become big because they have a great idea and figure out how to push that plan as far as it can go. Once they reach that point, it takes another big idea to get to the next level. However, a company built around one concept throws up all kinds of hurdles to and prejudices against things that are not part of the tradition of the company. Big organizations are also often siloed. Engineers and marketing people in one area of the business will integrate with resources above and below their own, but they may never talk to their counterparts in other divisions of the company. When new ideas require expertise or buy-in from other departments, turf wars and logistical problems can derail even the best proposals. Internal communication is a constant challenge.

A quick look at technology history demonstrates how rare it is for a large, successful company to start something entirely new. No big broadcaster started a cable company. No cable company started a satellite dish company. None of these companies created a freestanding Internet service.

Even recently successful, innovative companies have failed to successfully grab opportunities. Microsoft did not create Google. Google did not create Facebook. Facebook did not create Twitter or Groupon. Those companies almost certainly won't generate whatever comes next. Every day, especially in the Internet sphere, where barriers to entry are low, existing companies very rarely invent a compelling new service.

Executives at large companies are not stupid. They recognize that the bigger the company, the more difficult it is to start a new business. Half the world's corporate-retreat centers would go out of business if business leaders stopped trying to figure out how to break into new areas.

One solution favored by many large-company leaders is to create special teams of people whose sole function is to bring forth new ideas and make sure the ideas do not get killed by the corporate bureaucracy. To execute on new ideas, these "skunkworks" operations have their own personnel, financing, and incentives.

Another important strategy is growth by acquisition. If it's hard for large companies to innovate in-house, they can just outsource that function by scouting for and then acquiring smaller companies that are already active in an area identified for future expansion.

Companies typically go through a "make-or-buy" analysis to determine whether it is better and less costly in terms of money and time to develop new capabilities through acquisition rather than try to build a new operating business unit. The acquisition of Instagram by Facebook for $1 billion in 2012 reflected Facebook's desire to quickly enter the area of social photography.

Overall, growth by acquisition is far easier to describe than accomplish. Cisco is one company that has been masterful in this regard. Its CEO, John Chambers, has been especially skilled in the difficult art of integrating acquired companies into Cisco's corporate culture and operations.

It's not only a matter of size; sometimes it's a matter of being hungry or having a corporate culture that can embrace something new. Sometimes, successful company leaders are distracted by intense competition or challenges they face from government— which investigates and penalizes them for their success. This seems to happen more frequently in the United States each year.

The challenge of success is that it means everyone wants to go after you, and even though you may be number one, you feel like you're being attacked from all sides. There simply is not enough bandwidth among top executives for them to focus on expanding their businesses into totally new areas, especially when the financial community does not reward them for investing in products and services outside their traditional scope. Such investments are often viewed as coming directly off the bottom line and, moreover, likely to fail, given the track record of big-company innovations.

At CEA, I'm constantly worried about whether we're falling victim to the same hierarchical, stilted thinking that affects so many large organizations. How can we make the International CES more exciting year after year? Are we doing all we can to protect and advance our members? Where are our weaknesses, our flaws? What gaping holes do our competitors see that we do not?

To guard against these concerns, I employ an age-old tactic: the brainstorming session, which is simply a meeting where anyone can present an idea without judgment. Sometimes these ideas generate discussion for clarification, sometimes they're scrapped, and sometimes they produce even better ideas.

The beauty of a brainstorming session is that the participants usually are excited and energized by the process. They feel like they're part of the organization's success because their voices are heard. Also, they realize that no one has a monopoly on good ideas. They realize that ideas stem from a group of energized and committed colleagues who help build the respect and cohesion of the group. Most significantly, almost anyone can initiate or lead a brainstorming session. Recently we had a gap in senior management and the middle manager simply convened a brainstorming session to focus on a specific challenge. She had quietly and competently worked in the organization for many years and did not stand out. After the session, in which many top executives participated, I think many of us looked at her with greater respect because she convened and led a meeting that would likely be the pivot point for our success on a major project.

Brainstorming for us is a critical tool that ensures that we get the best ideas out on the table. Its success as a tool depends on the willingness of the participants to engage, present new ideas, and take risks. Brainstorming isn't the only way to keep your enterprise nimble and elastic, but it does foster a variety of qualities required of every ninja company: the ability to think creatively, an engagement with all employees, and a fundamental belief that we don't have it all figured out yet.

Ninjas Recognize the Status Quo Is Always Short-Term

CORPORATIONS MAY NOT BE PEOPLE, BUT THEY ARE MADE UP OF people—and people are funny. Our brains and our experiences tell us that things always change. Yet our instinct is always to preserve the status quo. We try to forget that we age, as do our parents and grandparents. The natural cycle of life is that we lose people we love, too often after barely taking the time to appreciate them in the present. We remember them as if they were alive and often regret the unasked questions and unspoken emotions. And of course in the future the cycle will repeat. We will lose others and perhaps have new regrets. But for some reason, we resist change rather than embrace the fact that the nature of life, of Earth, and of our condition is one of change. How we react to the change not only measures our adaptability but also determines our happiness.

Companies are exactly the same. I am always amazed how much corporate planning assumes the status quo. We will analyze market trends and look for opportunities to introduce new products or services, but then we forget that our competitors are looking at the same trends and considering the same decisions. In the consumer electronics space, any hot new trend is followed immediately by several competitors, each of whom sees the new trend and assumes no one else will. Dozens of new competitors with similar products often shake out to very few. Too many companies lose money by rolling out "me too" products rather than taking a risk, doing something different, or figuring out a variation that gives the product a unique selling proposition.

The tablet market is the best recent example. Any number of hardware manufacturers had introduced tablets, but none had caught on before Apple cracked the code with the original iPad in April 2010. Since that time, more than fifty companies have introduced or demonstrated tablets, most at the International CES. Yet Apple continues to dominate, and only a few competitors are still standing. Were the companies who ventured into the tablet market savvy ninjas or were they simply trying to chase a winner?

Given the inevitability of changes in the status quo, every company at some point has to make fundamental changes in strategy to survive. Once-great brands like Eastman Kodak, Circuit City, and Coleco all fell by the wayside after they couldn't figure out how to adapt to new realities.

Yet consider a couple of companies that fundamentally changed and are still household names:

Motorola started making car radios in the 1930s, during the height of the Great Depression (innovation can slow, but it never stops). It invented walkie-talkies, including the backpack model still famous today from so many World War II movies. It developed radios for NASA, including the system used by Neil Armstrong and Buzz Aldrin on the surface of the moon. As times changed, it branched out into set-top boxes, emergency communications systems, and of course, mobile phones, where it created two of the three most iconic modern handsets, the StarTAC and the Razr. (We'll give the iPhone top billing for now.)

Motorola focused on communications, but until recently it

had consistently managed to stay abreast of the competition by agilely identifying where technology and the market was headed. Motorola split into two companies in January 2011, with Google quickly snapping up the mobile handset side of the business in a $12.5 billion cash deal.

I mentioned IBM in chapter 1, but the company's storied history bears repeating here. IBM began as a "business machine" company and as technology evolved, it shifted to punch cards and huge mainframe computers. It dabbled in personal computers in the 1980s but never found a way to make enough money on the low-margin consumer market, despite some great innovations, including biometric control and the great center-keyboard mouse—a close friend to business flyers everywhere.

By the end of 1993, IBM was in deep trouble, having lost $16 billion in the three previous years. In an unusual move, the company hired an outside CEO, Louis Gerstner, who developed an entirely new direction for IBM as a high-margin service provider and systems integrator. The company has continued an aggressive move into software. By 2015, IBM expects half of its profits to come from that segment.[8]

Innovation by Design

IN A SPECTACULAR FORMER CHRISTIAN SCIENCE CHURCH OVER-looking Cleveland, the business innovation company Nottingham Spirk quietly develops new products for hundreds of consumer and medical product companies. Despite the fact that most people have never heard of them, Nottingham Spirk is behind many of the products we use every day.

John Nottingham and John Spirk are true ninja innovators. Since launching the firm in 1972, the two Johns have seen the company receive more than nine hundred patents through a development process that prioritizes openness, speed, and buildability.

The out-of-the-box ninja innovation of its founders is reflected in the company's location. Most businesses don't take up space in former churches. But Nottingham Spirk's leaders saw opportunity in an abandoned, aging, but architecturally significant church. Nottingham Spirk's leaders made an innovative proposal: They would acquire the church and renovate it into a world-class innovation center in exchange for historic-preservation tax credits to help fund the project. Nottingham Spirk brilliantly kept its word and renovated the church with awe-inspiring results.

Today, innovation at Nottingham Spirk occurs in a huge sanctuary with a soaring ceiling and a magnificent five-thousand-pipe organ. The engineering and extensive prototyping is accomplished in the former Sunday school on the lower levels. Consumers are often brought into the Innovation Center's insights lab, where needs are uncovered and the project team is inspired by their stories. The open circular plan and stacking of floors encourages communication and propels project momentum.

Similarly, the management structure is unconventionally flat, and job titles are irrelevant. Everyone, from the most senior designers and engineers to the newest college interns, works shoulder to shoulder in teams, and most serve on more than one team at a time. All are not only encouraged but expected to speak up, and all opinions are valued.

The whole process works because Nottingham Spirk's culture embraces the entire creative process, including the frustrating parts. No one is chastised for following a seemingly promising lead to a dead end. Experiments that fail and ideas that don't pan out are valued for the lessons they impart—lessons that may save time and money on future client engagements.

If this sounds too good to be true, consider the company's extraordinary forty-year track record. Odds are good that you've used one or more of the consumer products they've brought to market, including the Dirt Devil line, the Swiffer SweeperVac, the Spinbrush, the Axe pocket aerosol can, the Sherwin-Williams Twist & Pour plastic paint container, the Arm & Hammer Fridge Fresh, or the Scotts Snap fertilizer spreader. You may have encountered their specialty retail displays, like the personalized M&M printer or the Country Pure Chiller, the first countertop refrigeration unit. And if you haven't already, you may experience one of the company's medical devices, like the UroSense, the CardioInsight ECVUE vest, or the revolutionary HealthSpot Care4 Station, a sci-fi-like advance in convenient and cost-effective doctor-patient interaction.

Ninjas Play Chaos with Their Competition

I'LL CLOSE THIS CHAPTER WILL ONE LAST STORY FROM THE CONsumer electronics space. Twenty or so years ago, we were in the midst of an intense battle between two different camcorder formats. The two were running neck and neck in consumer sentiment, and the continual battle to produce the smallest, highest-quality camera was on.

That year, at the International CES in Las Vegas, Sanyo displayed an astoundingly tiny camcorder enclosed in glass at its exhibit. Throughout the four days of the show, crowds swarmed that transparent enclosure to take pictures of the amazing device. Competitors were blown away by something that represented a huge step forward in terms of engineering.

After the show, I asked a friend at Sanyo how in the world they had gotten their camera so small. His response: They hadn't. The "camcorder" didn't even work. Sanyo just wanted to mess with their competitors.

Simply by having an unexplained, impressive-looking prototype in a glass case, Sanyo became the talk of the show and threw product-development teams across the world into fits of consternation. Ninjas think outside the box—perplexing competitors through a head fake is both fair and clever.

CHAPTER EIGHT

INNOVATE OR DIE

THE NINJA GOES INTO BATTLE WITH FEW WEAPONS AND FEW RE-sources. He cannot rely on greater numbers or superior firepower because the enemy will have him beat in spades on both counts. He holds no hope for rescue or mercy if things go wrong and he is captured. Of all the rules that governed the use of ninjas in feudal Japan, one was paramount: Spies would not be tolerated. Each operation had only two outcomes: you either completed it or you died in the effort.

But the ninja was not without advantages. In a one-on-one fight, the ninja had the edge with superior training. If trapped, the ninja had the skills and the tools to escape. Above all, the ninja was a master of his surroundings. Only in very rare circumstances would a ninja find himself in a hopeless situation. He had the cunning, creativity, training, and tools to use anything and everything to his advantage. The ninja might make mistakes, but he would not be defeated by them. The ninja had no choice but to live by the words *innovate or die*.

Following the 2008 financial meltdown, "innovate or die" was the same simple message I had for my members. As an industry, we could either be dragged down with the rest of the economy or we could do what no industry does better: We could innovate. If we didn't, then we would have accepted that the situation was hopeless. I'm proud to say that almost all CEA members innovated, and, as I pointed out in the introduction, the U.S. economy in many ways has been propped up by the consumer electronics industry.

Admittedly, to observe that an industry like consumer electronics must innovate is a bit like saying the oil industry needs to produce gasoline. It's what we do. Our customers expect remarkable new products and ideas from us—and at a pace that few other industries can rival. We tend to forget, but only in the last twenty years or so has innovation in the CE industry proceeded at such a hectic pace. The VHS platform had a run of about thirty years before the mass adoption of the DVD player. That's nothing compared to the seventy years that the vinyl record dominated the music industry until the coming of the CD in the 1980s. As for printed books, well, I doubt they'll ever disappear—certainly not like the VHS tape—but Amazon's announcement in 2011 that its e-book sales had surpassed printed-book sales was surely a turning point.[1] Meanwhile, our lives are dominated by new technologies—smart phones, social networks, mobile apps—that did not even exist ten years ago.

But now, no one would place a bet that DVDs will last as long as the VHS tape. Indeed, DVD sales have declined for seven consecutive years, as consumers turn more to streaming platforms like Netflix.[2] The CD is already in a museum, and the product that helped put it there might soon be too. After peaking in 2008 with 22.7

million in sales, Apple's iPod has been on a downward trajectory.[3] Though that isn't so much a change in the format—digital music— but a shift in the preferred device (smart phones, for instance).

Which means that even in consumer electronics, the need to innovate is more necessary than ever. Manufacturers simply cannot be certain that a given platform or device will dominate the industry for longer than a few years before something else comes around. Twentieth-century economist Joseph Schumpeter famously called this process "creative destruction." You know the drill. Just when you've finally converted all of your home VHS tapes to DVD, you'll need to start the process all over again. It can be frustrating for consumers, and yet we gobble up the new devices, media, and platforms each and every year.

Throughout my tenure at CEA, I have watched dozens of companies dwindle and die—taking with them thousands of employees. Some of them were once powerful behemoths in their trade but became nearly forgotten footnotes in history. Changes in format, technology, and media can render any company—or even industry—obsolete. And while we should regret their loss, particularly the pain experienced by the now out-of-work employees, we cannot look back.

As for the companies that succeed, either as new start-ups or decades-old corporations, there is nothing magical about them. Which is not to say that what they do—or how they do it—is easy. The many successful companies profiled throughout this book— IBM, eBay, Amazon, etc.—have all stayed on top because they know that if they don't innovate, then they will die.

But the phrase "innovate or die" should not be exclusive to my industry. I believe it must be the national rallying cry to restore our economic prosperity. It shouldn't just be on the lips of every CEO in the United States. It ought to be on the lips of every member of Congress, heard in the halls of each federal agency, and repeated ad nauseam in the White House. It should be the rote Washington response to any company that calls to beg for a handout, a subsidy, an antitrust suit, or a self-serving bill.

Whenever the farmers call asking for yet another agricultural subsidy, the answer should be: "Innovate or die."

Whenever the electric car industry just needs a few hundred million dollars, the government should respond: "Innovate or die."

Whenever a steel company asks for tariffs on imported steel, it should be told: "Innovate or die."

What do we get instead? Each of the industries I just mentioned got what they wanted from Washington. The usual refrain from economically illiterate politicians is that this subsidy, that loan, or another tariff will "save jobs." It might—for that industry. But whenever Washington sticks its nose in the workings of the free market, there's always a loser. And the loser is usually the one who doesn't have the direct line to the committee chairman's office because they're too small to notice.

In today's political environment, "innovate or die" has been modified to "innovate or beg." The fact is that we've let it become too easy for companies and industries to cut a deal with government when their backs are up against the wall. The ninja company fights its way out through innovation, not backroom deals.

Even then, Washington has a tendency to discount innovation as a force for economic prosperity. President Obama has cited ATMs as a symbol of the negative consequences of automation. The thinking is that an activity that gave someone a job has been handed over to a machine. So, goes the logic, that's one less job in the economy.

Apparently, no one was hired to build those ATMs, which must have just magically appeared. And similarly, the erstwhile teller must have no other discernible skills other than cashing and depositing checks. It would be funny if this was not how Washington actually conducted its economic policies.

One should wonder what Washington will say when consumers start using ATM-inspired machines for their health care needs. This isn't a hypothetical. A start-up venture based in Ohio, HealthSpot, is reinventing the way we receive health care. Taking the idea of the ATM, which satisfies nearly all of your basic banking needs, HealthSpot has created kiosks called Care4 Stations that would do the same for your basic health care needs.

According to its creator, Steve Cashman, a thirty-six-year-old electrical engineer from Ohio, the Care4 Station could solve the problem of health care access for millions of people. Traditional health care requires travel, spending a lot of money, and waiting to get the valuable time of a doctor. With four children, Cashman realized just how much time he spent getting basic health care for his family. Combine that with Cashman's love for solving problems and his passion for technology, and you come up with an innovative idea. Cashman's solution solves this problem because it allows access to doctors anywhere.

It works like this: You go into a HealthSpot Care4 Station, which has a chair and video screen and some neat-looking fixtures. When you sit down your weight is taken. Through a few deft moves, your temperature, blood pressure, and other vitals can be recorded. You use the touch video screen to indicate any other symptoms. Soon you're linked up by video with a doctor, who will ask you questions and can unlock cabinets to get other medical devices that can help assess your condition. Within minutes, the doctor can issue a diagnosis and, if necessary, prescribe medicine. If the kiosk is in a drugstore, the prescription can be filled immediately.

This is a big ninja idea. It is out-of-the-box thinking (although in-the-box medicine). It satisfies a demand for easier access to care, it is cheaper than the average doctor visit, it would relieve overworked nurses and receptionists, it is clever, and it can work. Cashman has attracted considerable investment and plans to deploy more than a thousand Care4 Stations in 2013.

Should we criticize HealthSpot for taking the jobs of receptionists and other health care practitioners? That seems to be the answer coming out of Washington, if President Obama's views on ATMs is any indication. Or should we applaud a company whose product seeks to make health care more accessible for everyone? We need to rid ourselves of the notion that businesses exist only to provide jobs. If they did, then we would have millions of broom pushers, bank tellers, travel agents, town criers, wheat cutlers, and cotton pickers. Technology has replaced many of these jobs and consumers have benefited.

Businesses exist to provide a service or product people want—and that helps the larger economy. Their task is to continue to meet customer demands through innovative solutions. If they can't or refuse to, then propping them up through artificial means both delays the inevitable and hurts otherwise profitable companies and industries.

In other words, "innovate or die" isn't just sound advice for any company. It should be a rule that guides our national economic policies. Some might call it coldhearted. I'd counter that it's worse to sacrifice other companies or entire industries to prop up a dying venture. It's coldhearted to force consumers to pay more for something because of Washington policies that artificially raise prices.

Ninjas innovate. They might still die, but if they do, then they go down fighting.

The Bigger They Are, the Harder They Beg

THE AMERICAN BROADCASTING INDUSTRY, COMPOSED OF RADIO and television stations, is an example of a monopolist industry that long ago gave up innovating. But, by virtue of its government protection, it hasn't died—at least not yet. For years, TV and radio broadcasters had a monopoly on how Americans received their news and entertainment. They both played a monumental role in shaping not just American society, but our entire modern age. Then, starting about thirty-five years ago or so, the society they helped to forge began to change. And for the last several decades, the broadcasting industry has been dragged into accepting this change. And it's still kicking and screaming as loudly as ever.

But before criticizing the broadcasters, I should point out that they have a unique predicament. Both radio and TV broadcasters share a common legacy: They were loaned spectrum by the federal government to provide broadcast service in defined areas. Because spectrum is a public good, owned by the government and lent to broadcasters, the government requires that broadcasters operate in the "public interest." This vague standard empowers a federal agency, the FCC, to heavily regulate the industry and, thus, stifle innovation. Twice, when I was invited to speak to broadcasters at a meeting of their trade group, the National Association of Broadcasters (NAB), I asked the leadership if they would press a magic button to eliminate their regulations if they could. Their leaders responded, "No!" and explained that they thought regulation was fine. But later, in private, many individual broadcasters told me that these myriad rules were expensive and limited their ability to innovate and compete with other forms of media.

But while we can sympathize with the broadcasters, no one should shed a tear over the current plight of the broadcasters and their heavily regulated industry. The government strings certainly make innovation harder to achieve, but there are plenty of other industries that have similar responsibilities to the government, yet they manage to innovate all the same. With the broadcasters, it's a different story. They have accepted that the situation is hopeless, and they're always prepared to cut a deal with their captors.

First with the dawning of the cable age, then the satellite age and the Internet age, both the radio and TV broadcasting industries have lost market share to other media. In response to this unprecedented competition, the industry made several strategic mistakes, which only further eroded their market share. As monopolist purveyors of a "public good," the industry has never felt the need to truly innovate their business model to bring it up to speed with our new digital age. Instead, they have exploited their government ties not just to keep things the way they are, but to also actively turn back the clock to their monopolist heyday. The industry strategy is nothing more than relying on more government regulation and forceful lobbying so they can obtain or maintain a government-mandated advantage against other industries.

Take radio music. For years, radio broadcasters had it great. I mentioned earlier in this chapter how some platforms lasted a long time before they were replaced. Radio technology is a great example. It has not changed fundamentally since the turn of the last century, when Italian physicist Guglielmo Marconi realized that radio waves could be used to transmit signals the same way a wire can carry electricity. The reason is because there was no need (or really any way) to improve on the basic concept. Sure, the phonograph and vinyl records were a competitive alternative for listening to music, but only when one was stationary, such as listening in the house. The great advantage of radio was that you could take your music on the go. But the radio broadcasters had no other industry competition for portable music listening until perhaps the introduction of the portable audiocassette in the early 1970s. That's a long time to get fat and lazy.

Then things began to move rapidly—at least by radio industry standards. Audiocassettes made their way into cars, and next Sony blew away everyone with its Walkman in 1979. Then came the compact disc. The Walkman and its progeny offered portability, the cassette allowed customization, and the CD offered superior audio quality. By then, radio wasn't left with much. But it would get worse.

Today, consumers have satellite radio, smart phones, tablets, and even television as sources for music and news. Radio music is fading into oblivion. Indeed, the digital audio advertising giant TargetSpot found that nearly half of eighteen-to-twenty-four-year-olds (47 percent) spend less time listening to broadcast radio in 2012 than they did in 2011.[4] Not surprisingly, the steepest decline in broadcast radio listenership was among these "digital natives." Meanwhile, research from the NPD Group found 43 percent of the Internet population listened to digital radio broadcasts in 2011—up from 29 percent in 2009.

Rather than innovating and being ahead of technological change, the broadcasters have tried to use their legislative power to smother new competition. They unsuccessfully implored the FCC to stop low-power radio, on the laughable premise that community radio stations were a threat to large broadcasting corporations. They went to the federal government and delayed the merger of the XM and Sirius satellite radio companies on the schizophrenic premise that the new entity would simultaneously be a monopoly (with no competitors) and a competitive threat to broadcasters.

Now the radio industry has turned to government to impose a self-serving idea of mandating FM radio chips in all mobile phones. The stated reason is—of course—the "public good," because when a disaster strikes, as the broadcasters tell it, people will need radio to tell them what's going on. Forget that consumers have a pretty good idea of what's going on via smart phones, SMS messaging, weather apps, and other common channels of information. This is like requiring that every computer have a ballpoint pen on the keyboard.

In the 1980s and 1990s, I would occasionally be asked to address the radio industry. Time and again I would urge them to embrace digital local radio because the CD, and soon digital satellite radio, were coming, and radio shouldn't be the inferior choice for sound quality. I also pushed the radio industry to embrace a standard called the Radio Data System, or RDS, a feature that allows a visual display of words on a radio, like the name of the song and artist. While it became popular in Europe and elsewhere in the 1990s, U.S. radio broadcasters resisted it for two decades and only recently incorporated it into radio broadcasting.

Similarly, high-definition (HD) radio has been around for many years. Although it is now almost standard in every car radio, most radio broadcasters have embraced it reluctantly and refused to put any early investment dollars into promoting it. Instead, the radio broadcast industry has spent millions in Washington advocating for silly ideas like the FM radio chip and insisting that radio should be the only medium that can use copyrighted music without having to pay the record labels any royalties. This is absurd—for example, Internet-based Pandora, one of radio's competitors, pays about half of its revenue in royalties to record labels,[5] and digital satellite radio (SiriusXM) pays them 7.5 percent in royalties,[6] and yet radio broadcasters pay the labels zero. Why? Because when the stations had a monopoly on the consumer audience, they argued that the music they aired led to an increase in record sales—therefore, it was free promotion for the labels and artists. Today, there is no rationale for that argument.

In June 2012 I testified on these issues before the House Energy and Commerce Communications and Technology Subcommittee. In my prepared remarks I said:

> CEA opposes not only a mandate for FM chips in cell phones, but also opposes the broadcasters' current effort to require a government study of this issue. The marketplace has shown that Americans are perfectly capable of deciding for themselves what functions and features they want in their smart phones. Wasting taxpayer funds for something as absurd as an unnecessary mandate on innovation is the kind of special-interest-driven expenditure that frustrates average Americans.
>
> Clearly, broadcasters have lost their historic monopoly on music transmission and now exist in a more competitive environment. . . . The correct answer for broadcasters, however, is not to beg Congress to protect their historic business model. Instead, broadcasters must do what other industries do when faced with new market entrants—learn to compete smarter and harder.

I was thrilled to see that legislators from both parties were waking up to the fact that the broadcasters' requests for an FM-chip mandate and refusal to pay royalties to labels and artists had crossed the line from advocacy to attempted crony capitalism.

> For every new media device, there are more and better ways of getting content. It is a challenge for us, getting our content out there . . .
> —LESLIE MOONVES, PRESIDENT AND CEO, CBS COR-PORATION, 2011 PWC ANNUAL GLOBAL CEO SURVEY

Local TV broadcasters have also seen their market share decline. When cable was first introduced in the 1960s, the broadcasters ignored it because the market was too small (every new market starts out small). They did the same thing when satellite TV was introduced in the 1980s. One notable exception was the Hubbard family, who owned a bunch of Midwest broadcasters and invested heavily in starting a satellite company that eventually merged with DirecTV. But the Hubbards are the exception to the rule.

When high-definition television was discussed as a concept in the 1980s, the broadcasters, both local stations and networks, were smarter. They engaged in the process and worked hard in pushing the United States toward the best possible system. By January 2009, when President Obama made his first big presidential decision—to extend the deadline for the DTV transition—fewer than 10 percent of American homes relied on free over-the-air broadcasting. And sure enough, as we predicted, few complained or cared when analog broadcasting ended. The broadcasters had convinced President Obama and the Democratic Congress to waste over one billion dollars subsidizing converter boxes and delaying the transition.

Broadcasters have consistently opposed any innovation re- garding the way in which consumers receive their programming. Broadcasters sued to stop the technology we know as the DVR, first introduced at the 1999 International CES by the company ReplayTV. The suit was never resolved because ReplayTV filed for bankruptcy. In 2012, the broadcasters sued a follow-on product to the DVR, DISH Network's AutoHop, which allows users to skip over commercials.

Perhaps the worst stance the broadcasters have taken was to oppose a government-sponsored auction of the underused broad-band they owned. Wireless spectrum is the oxygen that sustains our mobile devices. The FCC estimates that smart phones consume 24 times as much data as traditional cell phones, while tablets can use as much as 122 times the data. Analysts forecast that mobile broadband traffic will increase 35-fold over the next 5 years.[7]

So we need to address the spectrum crunch. Americans rely on these miracle devices, and they are increasingly essential for the military, health care, education, and business. Government and the industry agree that additional spectrum is needed to avoid a devastating wireless traffic jam resulting in slower wireless speeds and unviewable video content in many U.S. cities.

Yet, the broadcasters just sat on their underused spectrum, waiting for the best possible deal from the government to sell off their borrowed spectrum to the wireless providers. Finally, in 2012, as part of a compromise over a larger bill, President Obama signed legislation that set up a wireless broadband auction handled by the FCC. The spectrum will be used by wireless broadband provid-ers as demands from smart phones and tablets have exceeded our wireless telecommunication system's ability to provide the fast, full-motion video that increasingly Americans want and rely on. For

TV broadcasters this is a great deal because those that sell will be paid for an expiring spectrum license.

Seeing that opposition was politically unsustainable, the broadcasters declared their public "support" for the auction legislation. Yet, the head of the NAB, former senator Gordon Smith, during his industry's annual trade show in March 2012, delivered a speech discouraging broadcasters from participating in the auction. In response, I wrote to Senator Smith:

> *I write to ask that you reconsider your public (and also private) posturing on the law allowing voluntary incentive spectrum auctions. Your speech at the NAB Show appeared to be rather discouraging to broadcaster participation in these auctions. . . .*
>
> *Recent statements discouraging participation in and support of these auctions are not only inconsistent with the goals of Congress, but also are not helpful to competition necessary for a successful and competitive auction.*

Obviously, in large part, the broadcasting industry lacks a strategy other than to lobby hard to protect the status quo. As history has proven, this tactic only works in the short term. It doesn't take long for yet another innovation to send the broadcasters scurrying back to Congress asking for another favor. Relying on government to protect you from competition is absurd. It's akin to the harm inflicted on children by today's trend that "everyone is a winner." Failure is what educates and encourages greater effort. The federal-government protection of broadcasters has made them weak, complacent, and entitled. If you talk to the broadcasters, they will ignore their decades of declining market share and describe how wonderful and special and worthy of unique government largesse they are.

I wonder whether if the broadcasters had to do it all over again, they would choose less government protection and regulation. Perhaps instead they would create their own competition and launch cable, satellite, and Internet distribution and content companies.

Of course, it is easy for me to observe the history of declining market share for radio and television broadcasters and point out their failures. But what about going forward?

So instead of just criticizing the obvious history of declining market share of broadcasters, I instead offer the industry a strategy:

1. COMPETE IN THE MARKETPLACE.

The industry should take advantage of its lower-cost structure as a strategic strength and "own" the local geographic area. Owning the local area means going beyond broadcast. CBS is cleverly heading in this direction with push e-mails offering special local deals, like Groupon, through its local affiliates. For their local affiliates, networks should create a business model where they set up the local station and share revenue both ways. Daily push e-mails with deals can also tease the schedule and expand both viewership and revenue.

The unique selling strength of a local broadcaster is not just selling advertising. It is also engaging the local community. Some businesses may want to just buy ads. But some individuals may want deeper public exposure, and will pay to present the local news and thereby become well-known local personalities

2. THINK BEYOND THIRTY MINUTES.

No law says everything must be on the half hour. All the research that led to the uniformity of the half-hour schedule occurred before the Internet gave us Twitter, Facebook, e-mails, text alerts, and other forms of competition for your shows. Broadcasters shouldn't be afraid to experiment and shake things up.

3. STOP RELYING ON WASHINGTON.

Finally, the industry as a whole must stop relying on Washington to protect it. Instead, it should work to free itself from the government strings that hinder its ability to compete with cable, satellite, and the Internet. Remove all the content restrictions, programming requirements, retransmission mandates, and costly requirements imposed by bureaucracy.

Above all, it's time for the broadcasting industry to accept the fact that it's not 1955 anymore, and it's time to enter the twenty-first century. If they cut their ties with the government, they might finally learn the truth behind the phrase "innovate or die."

Mr. Mulally Goes to Detroit

No matter how bad Ford Motor Company's problems are today, they aren't as bad as Boeing's were on September 12, 2001.

—ALAN MULALLY[8]

As Alan Mulally would soon discover, he was wrong. Ford's problems were worse.

In 2005, one of the most iconic brands the United States has ever produced was on the verge of declaring bankruptcy. The company's chairman and CEO, Bill Ford, had exhausted all of his options. After taking over the company from its spendthrift CEO Jacques Nasser in 2001, the great-grandson of Henry Ford had discovered that the problems affecting Ford Motor Company were beyond his capacity to fix. His quest for a savior CEO had failed. None of the exciting, innovative car executives Ford approached wanted the burden of resuscitating the dying American icon—at least, not without assuming both the chairman and CEO titles. Barring a last-minute miracle, Bill Ford was going to lose his family's company.

Alan Mulally was a top executive at Boeing. He had spent his entire career in the aeronautical industry, and most of that was spent at Boeing, the top U.S. manufacturer of commercial airplanes. Growing up, Mulally wanted to be an astronaut, but an eye exam that revealed he was colorblind dashed his hopes. So, Mulally figured if he couldn't go into space, at least he could build spaceships. But that plan changed once again when a friend advised him that the commercial jet industry held a brighter future than NASA. At Boeing, Mulally was on hand to save the company following the September 11 terrorist attack. For Mulally, it was an opportunity to restructure the corporate behemoth into a leaner, more efficient version of itself. After four years, Boeing was back on top, a feat that most attribute to Mulally's deft handling.

In 2005, after several Boeing CEOs were forced to resign, Mulally was assumed to be next in line. But the Pentagon was growing weary of the scandals plaguing Boeing CEOs and let it be known that it was in the company's best interest to find an outsider as CEO. Mulally was passed over.

And that's when Bill Ford pounced.

The history I just related comes from journalist Bryce Hoffman's superb telling of the story: *American Icon: Alan Mulally and the Fight to Save Ford Motor Company*. A reporter at the *Detroit News*, Hoffman had a front-row seat to these troubled years for Ford, when it looked like nothing could go right for the company. He documents the terrible corporate environment Mulally would inherit, showing in gruesome detail how Ford executives battled each other for an ever-shrinking slice of the pie. It was a situation that Bill Ford knew he couldn't handle, and so he did what a true ninja innovator would do: He brought someone in from the outside to shake things up.

To say that Mulally's arrival at Ford was looked upon with skepticism from the old "car guys" would be an understatement. During his first meeting with top Ford executives, one tried to school the neophyte on the ways of the car industry, saying, "The average car is made up of thousands of different parts, and they all have to work together flawlessly."

Unruffled, Mulally responded, "That's really interesting. The typical passenger jet has four million, and if just one of them fails, the whole thing can fall out of the sky. So I feel pretty comfortable with this."

But just because Mulally was familiar with mechanics didn't mean he could save Ford. Indeed, as Hoffman relates, Mulally almost didn't take the job because he wasn't sure Ford could be saved. The year Mulally arrived, 2006, Ford was headed to a $12.7 billion loss. The Ford family, anxiously watching their stock value plummet to $6 a share, pressed Ford constantly to do *something* to save their inheritance. That something was Alan Mulally.

Just four years later, Mulally had remade Ford to the point where it was able to report a $6.6 billion profit. More amazing, Mulally did it all without a bailout from the government, unlike his two American counterparts, Chrysler and General Motors.

Which is not to say that there weren't scary years for Mulally early on. Before taking the job, he had asked Bill Ford if he was prepared to do what was necessary to save his great-grandfather's company. Ford said he was. One can only wonder whether Ford imagined that Mulally would go on to mortgage all of Ford's assets (including its Blue Oval logo) in order to borrow the $23.6 billion he needed to begin his turnaround.

Mulally proceeded to follow all the concepts of ninja innovation I described in earlier chapters.

First, Mulally transformed Ford's noxious corporate atmosphere, where top executives were more interested in protecting their turf and blaming the other guy, into a truly collaborative environment. Taking a page out of his Boeing days, Mulally instituted weekly executive meetings, where division chiefs had to provide progress reports. "There was nowhere to hide," Mulally told Hoffman. Honesty would not be punished; ideas, any ideas, were valued; and, to drive home the point, he tied executive compensation to the success of the whole company, not just their individual departments.

Next, Mulally corralled all the disparate Ford divisions from around the world into one company. Instead of a Ford of Asia or a Ford of Europe, there was just Ford. And in his early days, Mulally put all his focus on rebuilding Ford's true home in North America.

Then Mulally focused on the car. It sounds ridiculous to non-industry ears. Don't all car companies focus on the car? Obviously, you're not from Detroit. The way the Big Three had operated for years was as the piggy bank for the United Auto Workers, a union whose lavish benefits for workers forced the automakers to put out cheap cars no one wanted. So Mulally got tough with the union, threatening to send all Ford manufacturing to Mexico if they didn't play ball. The result was that Ford started building cars that Americans actually wanted to buy.

Last but not least, Mulally redefined Ford as a technology company rather than a car company. Instead of talking horsepower and RPM, he focused on the fact that consumers want the latest, best, and most useful technology in the car. He even used the International CES to highlight Ford as a tech company. The press noticed and Ford's sales grew.

Before Mulally arrived, many believed Ford could not be saved. Many more thought he was crazy to refuse a government bailout. Cornered, cut off from reinforcements, low on supplies, Alan Mulally took the ninja way. He would innovate or die. He fought his way out. He innovated. And he saved Ford Motor Company.

CHAPTER NINE

AN ARMY OF NINJAS

UNTIL NOW, WE'VE MOSTLY BEEN EXAMINING THE SUCCESS OF single entities that exhibit ninja innovation characteristics: individuals, companies, organizations, and governments. But ninja innovators also mirror their historical counterparts in another key way: Their hard work enables others to revolutionize not just industries, but entire political structures and even societies.

Japan was steeped in war in the sixteenth and seventeenth centuries, and the ninjas played key roles in the conflicts. The Iga and Koga ninjas, led by the infamous Hattori Hanzo, played an integral part in Tokugawa Ieyasu's rise to power as a shogun in 1600, which marked the end of the Period of Warring States. One story in particular, retold in *Ninja: The Shadow Warrior,* by Joel Levy, stands out:

> *In one incident in 1600, the Koga ninja again came to Ieyasu's aid, perpetrating a classic ninjutsu ruse to help him escape a difficult situation. The powerful daimyo, on the verge of achieving ultimate victory, was threatened*

with a potentially lethal ambush, so his ninja guard cre-
ated a dummy replica of their lord, filled it with explo-
sives, and set it atop the Tokugawa carriage, which they
escorted as if nothing were amiss. When the enemies at-
tacked, the gunpowder was set off, killing the Koga ninja
escort but also the ambushers, giving the carriage carry-
ing the real Ieyasu precious time to escape.[1]

Ninjas' contributions during the uprisings enabled the Tokugawa family to come to power and establish two and a half centuries of peace and prosperity in feudal Japan under the Tokugawa shogunate.[2] At least in this case, ninjas played a decisive role in altering the course of history.

A Friend of Freedom

THE SAME RINGS TRUE FOR TODAY'S NINJA INNOVATORS. WHILE their work is intensely focused and personal, it has ramifications far beyond the boardroom or the balance sheet. Ninja innovators have created technological revolution, yes, but they have also empowered more widespread social change as well.

This is especially evident in the role social networking technologies played in the Arab Spring of 2011. Dictators survive on their ability to control the citizenry, and a large part of that control comes from the ability to manage the flow of information. But recent innovations in social networking, combined with the rise of mobile technology, have provided a way around the government firewalls, giving the repressed citizenry the ability to communicate and organize.

While recent political developments in these states don't bode well for long-lasting democratic reforms, the Arab Spring was fueled by the natural human yearning for self-determination. But it was only realized because social networking tools like Twitter and Facebook empowered citizens to convey information on smart phones, tablets, and computers working outside the government-controlled information paradigm. It was no coincidence that one of the leaders of the Cairo demonstrations was a mild-mannered Google software engineer named Wael Ghonim. We can contrast the "domino effect" of these popular uprisings with twentieth-century examples in Nazi Germany and the Soviet Union.

In both countries, popular uprisings were rare. Under Nazi control, there are few examples of occupied people in Europe rising up against their occupiers. The most famous one, the uprising in Warsaw in 1944, was brutally repressed, and the Poles didn't attempt another until the last day of the Cold War. For the Soviet Union and its satellite states, we have more examples—Hungary in 1956 and the so-called Prague Spring in 1968 being the two most famous examples. Yet in both cases, we didn't see a domino effect at all. They were isolated incidents, and if the planners believed that their actions would lead to a groundswell of support throughout the Soviet empire, they tragically were mistaken.

Of course, this isn't because the rest of Soviet Europe and beyond was happy with its Communist overlords. A large part can be attributed to the inability of these other citizens both to know what was going on outside the Communist propaganda machine and to organize themselves effectively. Either way, it was the paucity of information and communication that doomed any attempts at organizing.

Which is what makes the Arab Spring so unique. The mass

upheaval was far more organic in its beginnings than the Czech or Hungarian uprisings, which were carefully planned affairs. The original impetus to the regional uprisings was a young Tunisian merchant who set himself on fire to protest the oppressive regime in his country on December 17, 2010. Police tried to keep his death quiet, but a cell phone video of his funeral procession made its way online within hours, stirring up anger among the citizens of Tunisia, who would go on to overthrow their government.

While not technically part of the Arab Spring, the death of Neda Agha-Soltan in Iran in 2009 was similarly influential in stirring up protests against the government there. While attending a rally protesting the questionable reelection of Mahmoud Ahmadinejad, Neda was shot in the chest. Video of her death was uploaded to YouTube, and Neda instantly became an internationally known martyr. While difficult to watch, the graphic video touched a nerve, awakening a deep-seated, fundamental human need for freedom.

The effectiveness of social networking tools in antidictator uprisings is not just anecdotal. A September 2011 study by the Project on Information Technology and Political Islam at the University of Washington examined more than three million tweets and gigabytes of YouTube content. It found that social media "played a central role in shaping political debates in the Arab Spring. A spike in online revolutionary conversations often preceded major events on the ground. Social media helped spread democratic ideas across international borders."[3]

I had the great privilege of hosting a delegation from the Arab information and communications technology sector at the 2012 International CES in Las Vegas. Representatives from thirteen Arab nations came together to examine the effects of technology on the

Arab Spring and explore opportunities for the Arab IT sector to grow. The political revolutions opened up countless doors for new research and investment in the region that never would have been possible under the old regimes.

It's stunning to think that none of these tools existed even a decade ago, and the infrastructures upon which they're built are relatively recent developments as well. Though they had been in development since the mid-1940s, cell phones weren't widely accessible until the introduction of Code Division Multiple Access (CDMA) by Qualcomm in 1994—an idea that was itself nine years in the making—followed three years later by the introduction of the first Wi-Fi standards.

It can be argued that the Arab Spring eventually would have occurred without the availability of these technologies, but it definitely wouldn't have happened so quickly or visibly without broadband and mobile devices creating global awareness and enabling global support and encouragement for the spread of democracy in the Middle East.

The truth is that technology is often a friend of freedom. Before the advent of the Internet and social networking tools like the ones used during the Arab Spring, revolutionaries utilized radio broadcasts to inform and mobilize. During the Cold War, Radio Free Europe broadcasts originating in West Germany penetrated deep behind the Iron Curtain into Bulgaria, Czechoslovakia, Hungary, Poland, Romania, and even into the Soviet Union. The organization, which has since fittingly moved its headquarters to Prague, Czech Republic, now broadcasts into Iraq, Iran, Afghanistan, Pakistan, and other places where unfettered access to information is hard to come by, though new mobile technologies allow faster access to information and direct, person-to-person communication.

Politics by Other Means

To a certain extent, these technologies have engendered social and political change in the United States as well. On February 19, 2009, the video of CNBC commentator Rick Santelli's rant about the Homeowners Affordability and Stability Plan—a.k.a. the mortgage bailout—went viral online. Millions—far more than the number of people who watched it on the original broadcast on CNBC—saw it on YouTube and other sharing sites. The video is widely credited with being a launching point for the Tea Party movement, which organized over widespread frustration with government spending, deficits, and debt. The Tea Party was able to organize effectively and become a powerful political movement because it didn't rely on established Republican Party networks for funding and communication purposes. The Internet empowered them to gather information on their own and organize campaigns with minimal overhead.

While smaller and arguably less effective, the Occupy Wall Street (OWS) movement also finds its origins in social networking and mobile technology. By utilizing the Web's vast networking tools, OWS protestors were able to coalesce around their frustration with the disparity between what they dubbed the 99 percent and the 1 percent, not just in New York City's financial district, but in protests around America and indeed around the world. They used social networking to organize massive demonstrations and uploaded videos on their rallies to YouTube. Of course, for both the Tea Party and OWS, these videos—sometimes featuring the worst behavior in their ranks—were used against them as well.

Say what you will about the particular policies and practices of either the Tea Party or Occupy Wall Street, but the fact remains that neither movement would have been possible without the groundwork having been laid by ninja innovators. The tools created by the Mark Zuckerbergs and Jack Dorseys (of Twitter) of the world have made it possible for people's voices to be heard on issues they never would have been influential on previously.

One of the great miracles of innovation is that it breeds more innovation. Dorsey and his cohorts at Twitter set out to create a new way for people to share "a short burst of inconsequential information" with a small group of friends. But the phenomenon known as the free market had bigger ideas. Dorsey and his team never could have known their creation would fuel political movements in the United States, much less revolution in the Middle East. But it has.

While some are using these technologies to help them pursue political change, the rest of us are finding ways to use it to stay informed and share stories. When U.S troops entered Osama bin Laden's hideout in Pakistan in May 2011, the world first learned about it because a local Pakistani was tweeting about the commotion. Sohaib Athar became a worldwide celebrity for complaining about the noisy helicopters flying above his home in Abbottabad (who wouldn't?). Little did he know at the time that those helicopters carried the U.S. Navy SEALs who would take out the world's most notorious terrorist.

Innovation also helps save lives. When natural disasters strike around the world, including the devastating earthquake and tsunami in Japan or extreme floods in Thailand, mobile devices and social media enable rescue efforts to be targeted and effective. "Less than an hour after the quake [in Japan]," one report noted, "with the country's phone system knocked out, the number of tweets coming from Tokyo were topping 1,200 per minute, according to Tweet-o-Meter." Numerous media outlets set up Twitter accounts to post updates about the tragedy, while Google activated its Person Finder tool, which helps people reconnect with loved ones following disasters like the tsunami.[4] (Yet further evidence to contradict the radio broadcasters' claim that we need FM chips in our cell phones.)

More, innovation helps us heal by fostering community. When a gunman opened fire in a crowded movie theater in Aurora, Colorado, in July 2012, information was shared rapidly around the country and the world. In another famous example, a user of the website Reddit posted about his experience inside the theater, while others posted video from the scene. By the time the news crews got there, thousands of people already had a good idea of what had happened.

Almost immediately, other victims' stories started to come out via other networking platforms. We heard harrowing tales of how some brave moviegoers put themselves in the line of fire so that others might survive. As the hours and days wore on, more people shared their stories, whether on Reddit, Twitter, Facebook, or a host of other sites that connect people with one another. These public platforms allow us to tell our stories, process our feelings, comfort one another, and start to rebuild.

Innovation Builds Its Own Defenses

THERE'S ANOTHER IMPORTANT SIDE EFFECT TO ALL THE INTERCON-nectedness these innovations help foster. They make us, the consumers, stakeholders in their continuing success. Imagine the outcry if Uncle Sam attempted to censor Facebook or Twitter posts or impose taxes on their use. (Actually, you don't need to imagine it, as I'll explain shortly.) It wouldn't just be the companies themselves who stood to lose from government meddling; we'd all lose. In essence, these innovations have turned the masses into an army of ninjas, jealously guarding the tools that have led to a freedom explosion.

As powerful interests seek to stifle innovation, it's the people who are fighting back, advancing innovation through novel uses of the technological tools developed by the more recognizable, high-profile ninja innovators featured throughout this book.

In 2000, a team of ninja innovators created a music discovery service to help independent musicians find their audiences. One of the innovators, Tim Westergren, was a musician himself—he had played in rock bands and written film scores—and saw a need for technology that would solve the problem of discovery by helping listeners find new music and artists find new audiences.

Five years later, Pandora was unleashed on the world, letting users create their own Internet radio stations based on artists, albums, or genres they already like. Pandora picks the songs based on thousands of characteristics, greatly increasing the chance that the listener will hear something new that they like. The more feedback a user provides about the songs they hear, the smarter and more personalized their playlist becomes.

Pandora revolutionized the way we consume music and find new favorite artists, and has found great success. It is listed as the second-most-downloaded iPhone app of all time in the iTunes App Store, second only to Facebook. In June 2012, Pandora reported having a library of more than 900,000 tracks available to its 54.5 million active listeners, who accounted for 1.08 *billion* listener hours that month.

Pandora is increasingly prevalent everywhere. It is available in forty-eight vehicle models and hundreds of consumer electronics products, from computers and smart phones to Blu-ray players and Internet-enabled televisions. It is even built into some refrigerators.

But Pandora faces challenges. As mentioned in chapter 8, under some rather arcane laws, Pandora must pay half its revenue to performing artists, record labels, and songwriters. One of its main competitors, SiriusXM satellite radio, pays about 7.5 percent of revenue. Broadcast radio pays nothing to record labels. This imbalance is the work of a relatively small band of Washington lobbyists who have resisted every effort to require broadcast radio to start compensating artists and labels for airing their music.

There is no good reason why radio has such a competitive advantage. All music broadcasters, over-the-air, satellite, or online, should be operating on the same playing field. Instead, we have Pandora and satellite supporting the studios and artists while radio gets a break.

The only thing that will convince Congress to correct the lobbyist-created imbalance is a veritable army of American consumers. Pandora users must act to alert politicians that they care about musicians and Pandora and they want them both to succeed. Fortunately for Pandora and its millions of music-loving users, the American people have already proven they can and will defend innovation, often by using the very tools created by other ninja innovators in the process.

In 2011, the copyright lobby was pushing hard for legislation that would have allowed any copyright holder to shut down almost any innovative website by alleging a violation of intellectual property. Stopping this legislation, deceivingly called the Preventing Real Online Threats to Economic Creativity and Theft of Intellectual Property Act (shortened to the PROTECT IP Act, or PIPA) in the Senate and the Stop Online Piracy Act (SOPA) in the House, was a major priority for CEA, because it would have served to stifle online innovation—which is *the* future of all innovation.

The copyright lobby would not talk with us about fixing the bill, because they were convinced they could pass the legislation as they wrote it. They thought they could get exactly what they wanted in the legislation without having to compromise an inch. Indeed, the legislation passed unanimously through the Senate Judiciary Committee and had the support of most members of its House counterpart, but that was before the user of modern technology took notice. True to form, the copyright owners—the Hollywood studios and record labels—were living in the past. They failed to appreciate that the very technology they had spent decades opposing had changed the world.

As entrepreneurs, innovators, and website owners became aware of the legislation, innovation-friendly politicians like Representative Darrell Issa (R-California) and Senator Ron Wyden (D-Oregon) vocally opposed it. Then, as opposition to PIPA/SOPA built, online innovators staged a virtual strike. On January 18, 2012, many of the world's most popular websites—from the behemoths Google and Wikipedia to blogs and daily Web comics—shut down or ran statements announcing their opposition for that day to protest the proposed legislation even though their companies were fully functioning—a modern-day *Atlas Shrugged* as it were. It was annoying and frustrating for those of us who rely on the Internet to do our jobs or connect with friends and family, but that was the point. These innovators finally stepped out of the shadows to embrace their status as important—and yes, powerful—figures in our social and economic fabric, and demanded we start paying attention to their plight.

Many, like Wikipedia, the popular online encyclopedia, asked users to demand that Congress reject PIPA and SOPA and any other efforts to censor the Internet or stifle online innovation. Within twenty-four hours, Congress received some five million contacts from Americans angry about the proposed censorship, and more than thirty politicians withdrew their support from the legislation, killing it for 2012 and for the foreseeable future.

The people—this army of ninjas—not Congress, killed PIPA and SOPA. Leaders cannot govern without the consent of the governed, but in many cases historically they have been allowed to do so because of the barriers to civic engagement. In the pre-Twitter/Facebook days, it took a lot of time and resources to find out what was going on, spread the word, and rally the troops to oppose bad legislation. Technological innovation, however, has broken that cycle, making leaders accountable to the people all the time, not just every two, four, or six years at the ballot box. Welcome to the New Washington.

Pandora users need their own PIPA/SOPA moment. Unlike PIPA and SOPA, which targeted the entire Internet with overbearing regulations, the regulations that handcuff Pandora are narrowly focused, and really only affect Pandora and other similar services. The campaign will be difficult, but similar smaller-scale protests have been executed.

For example, on a hot Monday in July 2012, several thousand Internet users made a huge difference for one innovative start-up company that faced stiff opposition from the entrenched interests suddenly threatened by an up-and-coming challenger. Uber is a fast-growing service that connects smart phone users in major cities with nearby idle limo drivers. Many car services require three-hour minimums for chauffeured town cars or SUVs, so those vehicles end up sitting idle for long periods of time. Uber's summon-a-car app connects those drivers to users who want a cleaner, classier, and potentially quicker ride than they'd get from an on-demand limo service with its three-hour minimum. The service fills a huge gap between the randomness of taxis and the huge expense of limousines, and has taken off quickly in every city in which it has launched.

As you might imagine, the service threatens the entrenched taxicab monopoly, which went to work to shut down Uber. The D.C. Taxicab Commission was pushing for legislation that would have required sedan services—like the ones used by Uber—to charge at least five times more than a traditional taxicab. The bill even boldly admitted that it was intended to protect taxis from direct competition. What arrogance. But the CEO of Uber asked users to contact the city council, and in just a few hours, several thousand had responded. One councilman said his office received five thousand contacts in just a few hours. The proposed legislation was withdrawn and replaced with language that for several months specifically exempted mobile-phone-based sedan services from commission regulation.

Just a few years ago, such a stunning turnaround would not have been possible. Indeed, the Uber example is in many ways more remarkable than what happened during the PIPA/SOPA fight. This was just one company in one city. And yet, a Washington-based army of ninjas rose up to defend an innovative business model that fulfilled a demand that was not being met by existing services. The blatant pandering to the taxicab monopoly from members of the D.C. city council—which once would have gone on without any messy scrutiny—was laid bare for all to see. Humiliated and humbled, the council knew it had to change course.

And it's not just the people who have been recruited into this army of ninjas. Federal and state courts are also increasingly siding with innovation. An example occurred recently in New York City, where over-the-air television users often cannot receive signals thanks to the multitude of skyscrapers blocking the signals from homes. A service called Aereo was launched to help alleviate the problem by using a special antenna to resend the signal. It allows users to access live TV broadcasts at home or on the go through mobile devices and features DVR capability, all in an online interface that requires no downloads, setup, or special equipment for the user.

Broadcasters sued to stop Aereo from providing the service, claiming that the technology is illegal because it retransmits copyrighted content. Fortunately, a federal court in New York denied the preliminary injunction. The court based its decision on two other big court decisions regarding how consumers can use technology. The first was a Supreme Court verdict commonly known as the Betamax decision, which determined videocassette recorders (VCRs) were legal even though they were capable of recording full-length TV broadcasts—a particularly noteworthy episode in the history of innovation. The second case was a 2008 federal appellate court ruling known as the Cablevision decision, which found that a cable company's centrally located (as opposed to in-home) digital video recorders (DVRs) did not violate copyright law.

The Aereo decision reinforced the vitality of both the Betamax and Cablevision decisions as precedents that covered more than those specific brands. The court rejected the broadcasters' argument that those cases were limited to "time-shifting," where the consumer starts watching the recording after the broadcast is complete, strengthening the legal foundation in support of innovation that enhances users' access to broadcast television.

Similar challenges have recently been levied against new technology created by DISH Network that allows users to skip commercial breaks in programs they record on DVRs. Given the recent trends in courts finally starting to side with innovators over the entrenched special interests, however, things are looking promising for DISH and its consumers.

This movement for innovation also extends far beyond the world of consumer electronics, because the work of ninja innovators affects us all. Take, for example, a company called EHE International, which provides employee health and lifestyle management services. For one hundred years, the company has focused on providing preventive health care services, mostly as a perk companies provide to their executives. But recently, thanks in large part to technological innovations in the health care sector combined with a social and political environment that encourages rethinking our approach to health care, EHE has been expanding and growing. Its services are now directed at all employees—not just employers—because preventive medicine is the best way to stay healthy. While too many health providers focus on *treating* diseases, EHE is focused on *preventing* those diseases from happening in the first place.

Here is a company that not only saw an open door to promote a revolutionary approach to a major part of every person's life, but that also seized the innovations created by others to make that approach not only viable but attractive to consumers. Companies like EHE that don't necessarily deal directly in consumer electronics can benefit greatly from the innovation movement, so long as their leaders embrace the ninja approach to business: being creative, flexible, and adaptive, and capitalizing on the advances others have made and applying them to their own business models.

These kinds of people-powered changes, whether on the large scale, like efforts to stop SOPA and PIPA, or the small scale, like the outcries that allowed Uber to serve its users in the nation's capital or Aereo to allow TV viewers in New York better access to signals, are possible precisely because of the often-behind-the-scenes work of ninja innovators. They have allowed us to sidestep the filters of government and media, which too often serve the status quo, creating an information and networking revolution.

Sparking an Innovation Movement

INNOVATION HAS AFFECTED OUR CULTURE SO DEEPLY THAT IT HAS spurned its own sort of political and social party, known as the Innovation Movement. CEA started the Innovation Movement in 2009 after a survey we commissioned from Zogby International found that just 13 percent of Americans believed the United States would remain the world's innovation leader in ten years. More than a third said they expected the United States to take a backseat to China.

The Innovation Movement (DeclareInnovation.com) is built on the understanding that if we want a strong economy, Americans will have to work together to defend, revive, and promote innovation and entrepreneurship. Nearly three-quarters of Americans say that entrepreneurs who create and build companies are driving innovation today, compared to just 5 percent who credit policymakers who make spending and tax decisions. The goal of the Innovation Movement is to reverse the failed approach to economic revitalization that America has been pursuing for far too long and that too often demonizes business and entrepreneurship in favor of centralized control by the government.

There is long-simmering frustration on both sides of the political spectrum with government bailouts of failed companies, a practice that chokes out investment in new and innovative companies. People are waking up to the fact that overbearing regulations encourage companies to invest in jobs outside the United States rather than inside our borders. Citizens are recognizing that if we discourage the world's best and brightest from joining our ranks, they will find somewhere else to innovate.

For the United States to continue to lead the global economy, we need to pursue national policies that encourage innovation, creativity, and new ideas. We need to invest in technological innovation and create an environment where entrepreneurs can challenge, improve, and strengthen our society. Entrepreneurship, combined with technological innovation, will mend the global economy and lead us out of the economic doldrums. But just as ninja innovators have come to our aid, we must come to theirs.

The Innovation Movement is made up of regular Americans, businesspeople, and activists in both political parties who are putting aside traditional party labels, who are refusing to participate in the partisan bickering between left and right, and who are uniting around the innovative spirit that will save the American economy and restore our nation to its rightful place as the most productive, most prosperous nation on earth.

To date, the Innovation Movement has attracted more than two hundred thousand members whose lives have been touched by the consumer electronics industry, whether as active entrepreneurs and innovators or as consumers who use their products every single day. The grassroots campaign looks beyond party politics, sometimes siding with Republicans and sometimes siding with Democrats, whether on international trade, immigration policy, deficit reduction, broadband deployment, or a host of other issues that directly affect Americans' ability to innovate.

A central principle of innovation is that if the American people are free to choose their own ideas and pursue their opportunities, we can bring our economy back to life from the ground up. As we saw during the SOPA/PIPA debate, real change comes from below—when citizens, organized and informed through the very innovations a select few want to stifle, say enough is enough.

Thanks to the tireless work of ninja innovators in the past and present, we have the tools we need to get the job done. But, as an army of ninjas, we must seize the opportunity they have given us.

CHAPTER TEN

THE SHADOW WARRIOR

It's hard to believe that we've come this far without describing the one skill that clearly distinguished the ninja warrior, stealth. As a twentieth-century Japanese historian described it, "So-called *ninjutsu* techniques . . . have the aims of ensuring that one's opponent does not know of one's existence, and for which there was special training."[1] It was this ability to hide, to seemingly disappear, to surprise, that set the ninja apart from other warriors and martial arts fighters. In my tae kwon do experience, I learned how to punch, block, kick, balance, and maneuver, but I have no idea how to move about unseen. Yet this skill is the one that has survived the ages to define our culture's superficial understanding of ninjas. If you know nothing else about these ancient warriors, it's a good bet that you at least know that they operated in stealth—after all, what else was their black outfit for?

In reality, it's unlikely the ninja ever wore the all-black costume of Hollywood imaginations. Moreover, the ninja's stealth skills were not solely based on hiding. There are numerous accounts of ninjas operating in full view of their enemies, but in disguise or camou-flage. So while the ninja's stealth skills included staying hidden, they also included deception: tricking your opponent into not rec-ognizing a ruse. According to Barton Whaley, a leading expert of military-political deception, every deception is composed of dis-simulation and simulation; dissimulation is covert and hides or at least obscures the truth, whereas simulation is overt and presents a false picture.[2] The feudal Japanese ninja was a master at both dis-simulation and simulation. In either case, his enemies would not know he was there until it was too late. I have two favorite examples of such stealth, both quite different from my usual focus on private enterprise and applied-technology innovations.

Innovation Yesterday, but Not Today

MY FIRST EXAMPLE OCCURRED ALMOST SEVENTY YEARS AGO, ended World War II, and changed the world forever. It is the amazing story of the development of the atomic bomb, an un-precedented effort by the U.S military, civilian scientists and engi-neers, and many industrial firms from 1942 to 1945, code-named the Manhattan Project. Spread out at sites across the United States and eventually employing over a hundred thousand people, the massive project had to integrate diverse innovations in basic sci-ences, applied technologies, industrial processes, and aeronautics. Every site required very high levels of security and secrecy, but it was the effective concealment of the two main sites that best illustrates both ninja dissimulation and simulation. The primary scientific site in the bleak and remote desert town of Los Alamos, New Mexico, under the direction of J. Robert Oppenheimer, was built in record time from almost nothing into a small city covering fifty-four thousand acres and housing about six thousand scientists. The primary military site in the equally bleak and remote town of Wendover, Utah, was the world's largest gunnery and bombing range, covered 1.8 million acres, and was home to about twenty-five hundred airmen under the commands of two of the U.S. Army Air Forces' most accomplished officers, Colonel Paul Tibbets, who was responsible for the bombing missions, and Colonel Clifford Heflin, who led the base operations and ballistics testing of the bombs. As one example, all flights between the military airfields closest to Los Alamos and Wendover always landed at an interim military airfield before flying on to either of those destinations, which prevented flight plans from showing any direct connection between the two sites.

Although it later was discovered that spies for the Soviet Union (an uneasy ally at the time) stole our nuclear technology at Los Alamos, it took them four more years to test their first bomb, and our principal enemies during the war, Germany and Japan, never knew about, much less penetrated, either site. In fact, because all aspects of the project were smothered in secrecy and isolated from every other aspect, all but a handful of top military and civilian personnel ever knew what the endgame was. Even Vice President Harry Truman didn't learn about the project until he became president in April 1945, only four months before the bombs were dropped.

As sober and serious as the Manhattan Project success story is, my other favorite stealth example is lighthearted and somewhat playful. In the early 1960s, Walt Disney had his company secretly purchase forty-seven square miles of mid-Florida swampland where the only sign of human life was the nearby sleepy town of Orlando.

Disney had watched in shock and dismay as his first amusement park project, Disneyland, in Anaheim, California, was quickly surrounded and all but hidden among other businesses looking for a cut of the dollars being spent by hordes of tourists flocking to Disneyland. Walt vowed that before he built another amusement park, he would ensure that it would be a "world" unto itself. And, along with his brother Roy, who had a great head for business, Walt knew he had to forestall the possibility of land speculators learning of his plans and cheaply buying selected, well-placed tracts of land before Disney could get to them and then extorting exorbitant prices for the plots. So Disney, using a very small team, set up a host of dummy corporations to purchase the otherwise useless land, and by the time the *Orlando Sentinel* newspaper caught wind of what was going on, it was too late. Disney had successfully purchased enough land to form a buffer around his new park, which he named, appropriately, Walt Disney World. The first of several attractions opened in 1971, triggering the region's explosive growth, to the point that once-sleepy Orlando is now the most visited city in America.

The stories of the Manhattan Project and Disney World fit my view of the best of stealth innovation, which I will define in a moment. But I offer them first because I doubt either project would be quite as successful now. In today's world, stealth on these broad scales would have to contend with instant and ubiquitous communications, far fewer isolated locations, and seemingly less rigorous loyalties and ethics. In short, I think that attempted stealth innovation at the highest levels offers at least as many risks as opportunities.

Although stealth was a critical component of the ninja's skill set, I believe it is not as critical to successful innovation today.

Innovation Today, Maybe Not Tomorrow

I SUPPOSE EVERY HUMAN ORGANIZATION HAS SECRETS OF SOME sort, but not every organization attempts innovation. So, how should we understand and distinguish "stealth innovation"? Well, it has to start with distinguishing innovation itself, which several thinkers, beginning with Professor Clayton Christensen of Harvard Business School, have been analyzing for the last twenty years. Let's recall the three principal types of business innovation I mentioned in the introduction:

- Evolutionary: an improvement in an established market that competitors and customers generally *expect* to happen.
- Revolutionary: an improvement in an established market that competitors and customers generally *do not* expect to happen.
- Disruptive: an improvement that is generally unexpected by customers and competitors, serves a new set of customer values, and ultimately creates a new market that competitors scramble to understand and adapt to. If any innovation warrants stealth development, it's probably this type.[3]

Innovations can move between categories over time. For example, although faster computer chips may today be merely evolutionary, I would say the first computer chip was disruptive.

With these categories in mind, "stealth innovation" is when an enterprise gains an advantage on its competition with an innovation that is unexpected, creates a new market or an important new market segment, and shifts the innovation development focus of an entire industry. In these rare cases, stealth is a top priority because you want to lead the disruption, not be the victim of it. Just recognize that your innovation drives the need for stealth, not vice versa. There are far too many downsides to stealth maneuvers to attempt employing them with every innovation.

Unexpected and Unwelcome

IN MY THIRTY YEARS IN THE CONSUMER ELECTRONICS INDUSTRY, I've had a front-row seat to watch and contemplate the explosive growth and change of one of the world's most dynamic and innovative industries. To me, it's clear that, when not constrained by misguided government rules and regulations, highly talented businesspeople in most industries will regularly emerge to lead innovation, discovering new ways to better serve current customers and to create new customers. If the innovation begins to achieve some market success, with customers choosing it rather than competitive offerings, some competitors will react by matching the innovation or bettering it, whereas other competitors will find they can't keep up and instead look to other markets or simply fade away. Business casualties happen—via creative destruction—but in the best of circumstances, innovation begets more innovation and both customers and the industry in general benefit.

The danger facing every competitor in innovative industries, of course, is that although innovation in general is expected, it is not easy to predict and prepare for what is specifically coming next, especially in fast-changing, technology-based industries. Innovation is never dormant, but where free enterprise is allowed to flourish, innovators will thrive as well, and the pace and impact of innovation likely will increase. Think back to what I wrote about the dominance of the vinyl record or VCR. Both held on for decades before being supplanted by better technologies. Now no one expects most of today's electronic devices and platforms to last even to the end of *this* decade.

Even though many extremely bright minds are thinking about the future of the very same market, totally unexpected innovations happen. Sometimes it's because everyone is temporarily looking elsewhere; sometimes it's the result of an unexpected, serendipitous technical discovery; and sometimes it's because people just aren't paying close attention.

But occasionally it's because everyone has overlooked an innovation that was deliberately hidden until the very last moment before it hit the market. This doesn't happen very often, but that type of development strategy is talked about enough to have earned a special moniker: *stealth mode.*

Unfortunately, talk of innovating in stealth mode has come to mean one particular strategy: that nothing about the innovation is revealed until the moment it is introduced into its market, supposedly taking competitors completely by surprise, quickly capturing customers, and grabbing a dominant market share before competitors can fully react. It sounds terribly exciting, but in my experience that tactic is difficult, risky, and rare, so I'll try to provide a more balanced perspective.

Leakage, Now and Forever?

I WOULD BE REMISS IF I DIDN'T COMMENT ON THE NOTION THAT, FOR certain large firms and some federal agencies, stealth projects and programs are relatively easy to pursue and achieve. I believe that in this day and age, the U.S. government, much less a private company, wouldn't be able to keep secret for long a project whose scale was on par with that of the Manhattan Project. Stealth in any enterprise is exceedingly difficult to achieve, given all the moving parts inherent in any innovation worth keeping secret, not to mention the now-irresistible motivation to leak confidential information.

But while complete stealth is difficult to achieve, it is undoubtedly true that some level of secrecy and confidentiality is necessary in every industry, from birth facilities to mortuaries, from high tech to no tech, because every company has some amount of confidential information. And, as corporate espionage has become an increasingly urgent problem, it is almost impossible to overstate the need for thoroughly effective safeguards for such information. But at the same time, I believe that most innovative processes invariably will have leaks, with sources ranging from incautious employees to vulnerable communication systems, to the need for primary market research with potential customers to help guide the direction of innovation.

In addition, each of the various types of innovation presents other, unique security challenges. For example, technology development itself, the basic application of science to solve problems, tends to thrive more in environments that are open to contrary opinions, which are difficult to generate solely within a single firm. Similarly, new products, new services, new designs, and even new business models may require outside vendors to be involved in order to become 100 percent complete, and those vendors might have to be engaged earlier rather than later in the innovation process. In fact, although a firm's particular innovation process might be proprietary, participating scientists, engineers, marketers, and executives change jobs every day and bring their learning with them, regardless of nondisclosure agreements.

In this regard, I'd like to venture a cautionary observation about Apple's innovation process, which may be the best in the tech industry. As reported by Walter Isaacson in his recent biography of Steve Jobs, it's apparently true that Apple has a top-secret inner sanctum, with tinted windows, a heavy-duty door, and two guards; it's an office presided over by Jonathan Ive, Apple's brilliant senior vice president of industrial design, and holds a pipeline of several years' worth of potential new products that Jobs himself worked on.[4] But I often read Apple's annual reports, and one particular statement recently caught my attention:

"The Company's business strategy leverages its *unique* [emphasis added] ability to design and develop its own operating systems, hardware, application software, and services to provide its customers new products and solutions with superior ease-of-use, seamless integration, and innovative design."[5]

In a 2012 interview, Mr. Ive elaborated on Apple's innovation process when he said, "We try to develop products that seem somehow inevitable, that leave you with the sense that that's the only possible solution that makes sense."[6] This sense of the inevitable strikes me as being an important ingredient in Apple's "secret sauce" for success, yet I can't help but wonder if part of that recipe walks out the door every night. Indeed, in 2010, the then-secret iPhone 4 was found left at a bar, disguised to look like a previous model. The ruse didn't work.[7]

In any event, for all the above reasons, I'm skeptical that it's possible to maintain a totally secret stealth mode up until the moment of releasing an innovation and then also go on to thoroughly defeat your competitors within days, weeks, or even several months. But I do believe in more modest versions of stealth mode innovation, including for new products, services, and business models. The breadth of possibilities is too broad to catalog, but I'll offer some examples.

Amazon, Once More

PRETTY MUCH EVERYTHING I COULD SAY IN THIS BOOK ABOUT JEFF Bezos and Amazon has been said in other chapters, but you won't be surprised that, as I write this, Amazon is actively pursuing what I'll call a "semi-stealthy" innovation. According to a report in the *Financial Times* (London), Amazon is steadily expanding its U.S. warehouse network in order to put product inventories close enough to large markets to provide same-day delivery.[8] I describe this development as semi-stealthy because at the time I'm writing, I'm not aware that the company has made any formal announcement of such a strategy. Yet the story quotes an Amazon competitor as saying, "Amazon's business model has changed from being a remote seller without a physical presence in most states to a company that—through distribution centres and delivery lockers and the things it's doing to get close to customers—has a physical presence in lots of places."

Why this news strikes me as a particularly good example of stealth used properly is because Amazon's competitors—namely, any e-commerce merchant these days—seem to be stuck in a constant waiting pattern to learn what Amazon is going to do next; then they mimic it. With its development of Amazon Prime—a subscription service that provides users with more discounts and expedited delivery—Amazon pushed the rest of the e-commerce world to embrace next-day delivery. Now, because of Amazon, e-commerce merchants have to answer this question from consumers: "Why can't you deliver it tomorrow? Amazon can." It must be frustrating at the very least.

The *Financial Times* article ups the ante because it appears as if Amazon might be moving to a nonsubscription next-day delivery model, which puts further stress on its competitors. Just when e-commerce retailers are putting in place their own VIP next-day packages, Amazon is moving beyond them yet again. Or are they? Remember, stealth isn't just about keeping something hidden; it's also about keeping your competitors guessing about what they think they're seeing and in what direction you're actually going.

Amazon is America's, and the world's, "horse of many colors," to borrow an apt description from an iconic movie, and it has been adroitly displaying each of its many colors, for the most part, very successfully. Yet, as I write this, there are repeated questions about Amazon's ability to generate consistent profits in the face of the costs of innovating in so many different ways. But I'm not an investment adviser.

Mobile Mania

WITH ALL THE DYNAMISM OF TECHNOLOGY-DRIVEN INDUSTRIES, many have decades-long stories of growth, retrenchment, acquisitions, divestitures, triumphs, failures, and whirlwind changes in management. But to my mind, nothing quite illustrates the repeated impact of stealth innovation better than the story of mobile telephones. And that tale cannot be told without beginning with two of the CEA's longtime members, AT&T Inc. and Motorola Mobility Inc. Although the first "mobile" telephone was actually a telephone installed in an automobile back in 1946, I tend to date the beginning to 1973, when the first *handheld* mobile phone was produced by Motorola, which had been in a stealthy development race with the Bell Laboratories unit of AT&T. The breakthrough was technically exciting but commercially stymied, not least because mobile phones of course require wireless networks over which to operate—and those networks were few and far between.

Then, in the 1980s, two major events occurred that began to open the market: The U.S. Justice Department broke up AT&T by limiting it to handling long-distance calls and divesting its local telephone operations into seven independent companies, which were nicknamed Baby Bells; and almost simultaneously the Federal Communications Commission limited to two the number of wireless network providers in each city, with each Baby Bell automatically receiving one of the licenses for the cities in its region. Now the Baby Bells (and other, non-Bell local telcos) had the incentive to build out their networks and begin to look for ways to create a truly nationwide system. The process was messy and required a lot of stealth horse-trading to bring our mobile system to the point where it is today.

Additionally, the Baby Bells, having been part of a monopoly for virtually all their lives, were still learning what it took to successfully manage their semi-new companies in an increasingly competitive market. Encouraging top management to forget about the comfortable world of monopoly and learn about the frightening world of competition took a bit of finesse and skill—some might call it stealth. Such is the rigid "thinking" of monopolists.

Finally, after much strategic and operational turmoil, the networks grew to support reasonably user-friendly mobile phone service, and then the market began to really take off. Mobile devices have evolved into today's smart phones in tandem with the evolution of wireless networks, which went from analog cellular (known as 1G) to digital cellular (2G) to mobile broadband data (3G) to mobile ultrabroadband Internet access (4G) to . . . well, I won't try to predict what's coming next except to say that it will probably be exciting. Along with these technical advances, several companies that were once part of the old AT&T were reacquired, and in the process probably recaptured operational efficiencies that shouldn't have been broken up in the first place. In other changes, whereas once almost all U.S. telephones were made by another unit of AT&T, there are now more than a dozen mobile phone manufacturers, still including Motorola Mobility. And in early 2012, Apple became the third-largest mobile phone manufacturer in the world after only five years in the market.

Looking back, among the things I like about this story is that it shows that the spirit of stealth innovation need not be limited to development of products, services, and business models, but also can be useful in mergers and acquisitions, and in gently educating older generations about new technological realities. Our wireless network—though far from ideal—grew organically, but it was also the result of all that grunt work and stealth moves to get it off the ground.

Stealth Start-ups

EARLY-STAGE COMPANIES FACE SPECIAL DEVELOPMENTAL ISSUES that stealth mode seems to conflict with. For example, stealth mode, with all its secrecy, makes it more difficult for a new, unknown company to attract top-level employees and strategic partners; to arrange for beta and reference customers; to secure funding from professional sources (angels, venture capital firms, and corporate venture funds); and to build potentially vital networks. And from what I've seen lately, stealth mode doesn't seem to be as popular among start-ups as it once was. In any event, a while back I came across a story about the firm Transmeta, and it seems to have positive and negative lessons for stealth mode start-ups.

Transmeta was organized and began developing a low-power computer processing chip around 1985. It managed to stay mostly under the radar until mid-1997, when it put up a website with only this short statement: "This website is not yet here." This created a stir in the start-up world and apparently brought Transmeta some greater attention, although little or nothing was yet revealed about their technology. Then, in November 1999, the company posted another message on its minimal website, but this one said quite a lot:

> *Yes, there is a secret message, and this is it: Transmeta's policy has been to remain silent about its plans until it had something to demonstrate to the world. On January 19, 2000, Transmeta is going to announce and demonstrate what Crusoe processors can do.*
>
> *Simultaneously, all of the details will go up on this Web site for everyone on the Internet to see. Crusoe will be cool hardware and software for mobile applications. Crusoe will be unconventional, which is why we wanted to let you know in advance to come look at the entire Web site in January, so that you can get the full story and have access to all of the real details as soon as they are available.*[9]

I'm no expert in computer chips, but Transmeta apparently had something that promised to be very special, because only five months after that January 2000 information release, it raised $88 million in private funding, and seven months after that (one year after revealing its chip) Transmeta went public in a $273 million offering, despite having achieved only $5 million in sales the previous year.[10] It certainly seemed like the company's stealth mode strategy was paying off. But that's not the end of the story.

From that point on, things got complicated, so I'll try to summarize it by saying Transmeta's chips proved to not be as good as expected and apparently never achieved profitability; the company later restructured into being an intellectual property company selling technology to other chip makers. In late 2008, Transmeta was acquired by a digital video processor company called Novafora. In early 2009, Transmeta's patent portfolio was acquired by an intellectual property company, and in July 2009 Novafora went out of business. I suspect that operating in stealth mode for its first five years may have contributed to Transmeta's problems (e.g., did it sufficiently vet its technology with outside experts?), but I'll never know for sure.

In the end, Transmeta both profited handsomely and suffered greatly from its stealth skills. Had the chip performed as well as the sneaky advertising campaign, then maybe we would be admiring how Transmeta surpassed Intel as the industry's top chip manufacturer. But Transmeta seems to have grievously erred in thinking that stealth itself is a substitute for a compelling product. As I said earlier, any decision to wrap your innovation inside a stealth-level package first requires that your innovation be irresistible to customers.

But compared to many other stealth-obsessed firms, Transmeta was fairly successful. I recently came across a press release from a company I hadn't heard of that has a technology I don't entirely fathom. Nevertheless, the press release seems to well illustrate the fixation on stealth packaging to the detriment of clear thinking about your innovation and about smart communication with important constituencies. The company will remain anonymous and I've summarized the press release into what I hope are readily readable bullets:

- Headline: "[COMPANY] comes out of stealth mode."
- The product has been in development for eighteen months and "came through the research & development efforts of serial entrepreneurs, technologists, and co-founders."
- "The product enables users to mashup data from any source, slice and dice data, build ad hoc reports, publish dynamic dashboards, and collaborate among stake holders to make informed decisions."
- "The venture has flown under the radar until this week when [CEO] gained an angel round of funding to quickly expand [COMPANY] and its market reach."
- "We were tasked with a very complex challenge of innovating a cloud based analytics solution that can be used by everyone, must be implemented in less than 30 days, affordable to everyone, yet scale for Big Data . . . [so we] had to use a radical approach to design an innovative instant analytics technology stack that can support high volume data mash ups from any source, provide real-time responses to ad-hoc queries and scale massively and horizontally in commodity hardware."[11]

The above is what the company's communications team chose to highlight in the press release. It's not until the last two paragraphs that non-techie readers (including potential investors) get a sense of what's actually going on there:

> Businesses of all sizes are now competing in a world flooded with information. The winners will be the ones that can make the most sense from all the data mashed up together . . . and do it the fastest . . . Analytics is not just for large companies anymore. Small companies need to analyze data too . . . it is imperative to implement a metrics based management system, whereby proactive decisions are made based on meaningful and relevant data. [COMPANY] provides the platform and foundation through a plug and play analytics solution that can be implemented rapidly at a fraction of the cost . . . is ideal for any organization, department, analyst that wants to manage performance and optimize operations through business analytics.

Finally, we get a sense of the problem the firm is trying to solve and why its innovation may offer customers a unique value compared to competitive solutions. Again, I'm not offering any judgments about the company, its overall team, or its innovation. I merely want to point out that the writer thought it important to first say the company had been in stealth mode development; that the writer apparently thought his primary audience would be composed of analyst-level workers and not other important constituencies such as financial executives and potential investors; that the writer seems to believe it would be compelling to know the company has been flying under the radar, in stealth mode; and so on. It seems to me that the writer is so excited to report on the company's stealth mode and all the wondrous things its product can do, she's apparently forgotten the firm's elevator pitch, the short statement that explains why its product is a "must have." As Abraham Lincoln once said about a certain speaker, "[he could] compress the most words into the smallest ideas of any man I ever met."[12]

To Stealth or Not to Stealth

THE ABILITY TO OPERATE IN STEALTH WAS AN ESSENTIAL SKILL FOR the ancient ninja. It all but defined his status as a superior class of warrior. But my purpose in this chapter was to show why stealth mode development, although potentially valuable, is difficult to undertake, not necessarily an essential skill for a ninja innovator, and a potential snare for the unwary. Stealth is a form of secrecy, but not all appropriate secrecy requires stealth, and not all innovations will benefit from stealth. Moreover, although I have no empirical data to prove it, I believe the most successful innovators, like the most successful ninjas, develop a sixth sense, an instinct if you will, about the level of secrecy and stealth that fosters the best result, measured of course by success against the competition. They understand that the strength of innovation is, almost by definition, a product of its times.

You may notice that I do not suggest relying on our patent system as an alternative, or strong complement, to secrecy and stealth, first because I'm neither a patent lawyer nor a patent expert, but also because I'm increasingly worried that our system is weakening, if not already seriously broken. To be clear, I have no doubt patents and patent applications can be solid protection, but I am not able to describe how one can know this with certainty in advance, before the inevitable challenges and horse-trading have run their course. Maybe I'm unduly pessimistic.

In the end, I lean toward Peter Drucker's view that your company's primary purpose is to create customers, and that this is primarily driven by good general management, marketing, and innovation. He didn't seem to worry much about doing it stealthily.

EPILOGUE

THE INTERNATIONAL CES
KILLER STRATEGY

I TOOK OVER AS HEAD OF THE CONSUMER ELECTRONICS ASSOCIA-
tion in the midst of the 1990–91 recession. My immediate challenge
was that our biggest revenue source, the Consumer Electronics
Show, now called the International CES, was losing the popularity war to the COMDEX (Computer Dealers' Exhibition) trade
show. I complained to one of our key board members, the late John
McDonald, then president of Casio, that it was poor timing for me
to start out my CEO career staring down the twin barrels of the
recession and the COMDEX threat. He responded with his typical delightful and great wisdom: "Gary, any idiot can be in charge
when times are good. It takes someone with brains and moxie to
lead in tough times." Little did I know how difficult the path ahead
would be.

Back then, COMDEX seemed to have all the advantages: a huge amount of buzz, a fast-growing IT industry with lots of new and growing companies, a dedicated audience of geeks proud of their COMDEX credentials, and an entrepreneurial founder, the legendary Sheldon Adelson (who built and still runs the Venetian hotel and Sands Corporation). More, COMDEX was always held right before Thanksgiving, fewer than sixty days before CES in Las Vegas. In the trade show world, like in many endeavors, it is sometimes better to be first.

But the COMDEX owners weren't content to stick with IT—they were coming after us. They started courting consumer electronics aficionados and seducing our largest customers. We began hearing from companies who said that COMDEX had become more important because it had many more attendees and greater momentum than CES. In fact one of our largest exhibitors expanded its presence at COMDEX, and reduced its exhibit at CES, simply by saying COMDEX attracted more attendees. Never mind that they were essentially a consumer show, open to the public, and we were a trade show, only open to the people with a business interest in consumer electronics. In this game, it's all about numbers.

Our challenge wasn't just to beat COMDEX; in a tough economy, we were concerned about surviving. So over the course of the early 1990s, we developed a strategy. Today, I view it as a killer ninja strategy, perhaps because at the same time we were building up CES, I was also studying tae kwon do and going through the various tests to advance through each color belt on my way to a black belt.

Mistakes, We Made a Few

BUT BEFORE GETTING TO WHAT WE DID RIGHT, LET'S EXAMINE what we did wrong. We need to back up a bit and provide some context. The Consumer Electronics Show began in 1967. Prior to that, consumer electronics products were exhibited as part of the National Association of Music Merchants show. But my predecessor, Jack Wayman, thought that TVs, radios, and phonographs were not getting their due, buried, as it were, among aisles of musical instruments. He asked the association board if he could start a show.

The board did not immediately agree. Zenith and RCA argued that having a trade show for consumer electronics would allow new foreign competitors like Sony and Panasonic to have easy access to U.S. distribution. But Jack wore them down, and as I briefly noted in the introduction, the first CES was held at the Americana and Hilton Hotels that June in midtown Manhattan. Jack's instinct proved right. More than two hundred companies exhibited and an estimated seventeen thousand people attended—not bad for a first-time event. Over the next few years, the show grew quickly and in 1972 shifted to Chicago. In 1973 CES became a biannual event, with a winter CES and a summer CES, both held in Chicago.

Then, in January 1977, disaster struck. It was so cold that the winter show attendees could not leave their hotel rooms. That prompted another radical but huge decision. Jack and his team decided to move the winter show to Las Vegas. Some were concerned that Las Vegas, with its "Sin City" reputation, was inappropriate for a business event. But these objectors missed the point—Las Vegas was and remains a business-friendly location, because it is the only city in the world primarily focused on a great visitor experience.

The whole place is geared toward giving visitors a great two-or-three-day experience.

Of course the move west turned out to be brilliant. With our first winter show in Las Vegas we doubled our attendance numbers and exhibit area. From then on the Las Vegas show had steady growth and the Chicago summer show began a slow decline. By 1991, when I took the reins, the Las Vegas show was by far more important than the summer event. Indeed, one clever journalist dubbed the summer show in Chicago "the wake on the lake." It stung, but it was true.

But the city of Chicago wasn't helping at all, either. The unions were difficult to work with and many exhibitors started to complain. They especially did not like paying electricians to watch them install their own products, not to mention the huge price the unions charged for "drayage" (getting your exhibit from and to the building loading dock). The unions were also a problem for us as a show producer. For example, just to open the theater inside the convention center for a one-hour speech required a six-figure expenditure *and* we had to use several unions. In response, many exhibitors took the ethical low road. They wouldn't officially participate in the Chicago show; instead they would "outboard" and show products at nearby hotels rather than deal with the unions.

In desperation, we asked the unions to work with us ("help us help you"). It was in no one's interest to kill the CES, but the unions' actions were doing just that. I asked them to relax their overtime rules and base them on hours per week—rather than hours in a day—because setup and takedown are short spurts of time. Teamster head William "Billy" Hogan Jr. told me—typically—I was trying to take "food out of union workers' mouths." I tried to explain that we could not survive in Chicago and there would be no work at all if something did not change. I got nowhere. (As an aside, I did get a call a few years later from Hogan asking me to be a character witness for him in a trial. I politely declined.)

To keep the show relevant, the CEA board wanted to open the show to the public. The decision was made. It was a big change because the exhibitors were accustomed to dealing only with business executives coming to their booths. Most companies viewed it as an opportunity to showcase to consumers their products in a way that retailers did not. Although it was quite an effort to attract regular consumers and to educate the exhibitors on how to change their sales pitch, it did boost attendance at the Chicago show. Still, we couldn't change the unions' stubbornness with regard to charging exorbitant fees, nor could we change the fact that the winter show in Las Vegas was bigger and cheaper. In hindsight, opening the Chicago show to the public was a Band-Aid to a gushing chest wound.

Ironically, everyone in the trade show world thought our decision was revolutionary. They said CEA was starting a new wave of public trade shows and that all shows would go that way. It was a trend story, and like lots of trend pieces it was wrong. We were just trying something in desperation. The bookkeeping does not lie: Even if we had fifty thousand members of the public attend and each exhibitor sold to 1 percent of the customers, five hundred consumers buying your product does not justify the cost of exhibiting.

It was around this time that we made another bad decision. The video-game market was growing rapidly, and while we allowed those companies to exhibit, we didn't do the hard work of building lasting relationships with them. This was a huge mistake. When I suggested to the board that we do some things to help this industry because we were in danger of losing them, I met stiff resistance. Indeed our most vocal board member, John McDonald, was quite blunt: "F—— them," he said. "They will come to us on their knees." I responded almost tearfully that they were providing us $10 million in revenue a year, and over a ten-year period, even if they didn't grow, that would mean $100 million we would be losing. I lost the battle at the board that day. But my concern would be validated when the video gamers eventually left to start their own highly successful show, the Electronic Entertainment Expo, or E3.

The end to our Chicago relationship came when the city insisted that a major soccer tournament be held over our show dates. By then we had had enough. It was apparent that Chicago lawmakers were taking us for granted. And not only CES, but every industry that showcased in Chicago. So we moved the summer show to Orlando. But few people came—or at least chose the show floor over the Orlando golf courses and theme parks. And while the exhibitors appreciated the relatively low-cost, friendly Orlando labor, the writing was on the wall for the summer show.

We even tried combining our show with the COMDEX spring event. Boy, that was an even bigger disaster: Our corporate cultures clashed, and despite elaborate revenue-sharing agreements, there were constant battles, from how to split mutual customers to how we reported (or didn't report) facts like attendance and who was confirmed to exhibit. My lesson there was that you need chemistry between partners to make something work. More, two bad shows combined do not make one good show.

After that, we let the summer show die. We began a beautiful event in Mexico and focused on building our Las Vegas event.

For me, the death of our summer event was painful. I felt I'd failed. I know that had I done things differently we still would have video games under our tent. But I learned a valuable, if seemingly obvious, lesson: You must pay attention to your customers. We took the video gamers for granted, just as Chicago took us for granted. I also learned that you have to be careful about your old customers blocking your ability to court new customers. Since then, our board has never (unlike many association boards) prevented us in any way from going after new exhibitors. They recognize that we all benefit when the industry is growing and every segment is able to shine.

The World's Largest Consumer Technology Show

TODAY, THE WORLD OF INNOVATION FOCUSES ON LAS VEGAS FOR the International CES, the world's largest and most important consumer technology trade show. Every year, more than 3,000 companies display their best ideas to more than 150,000 visitors—reporters, buyers, investors, and potential business partners—in several cavernous exhibit halls and hotels. No matter the reason, everyone who attends the International CES has chosen to invest precious time and money. Their return is more than entertainment; it is business—and an awesome and inspiring display of the future.

They travel great distances to meet like-minded individuals and discuss the vibrancy, hope, and promise of innovation. They come because in no other venue can a company make a profoundly physical statement to five thousand reporters and analysts—and more than thirty thousand international business visitors. Yet the distance any one attendee travels to CES pales in comparison to the miles they would have to cover to accomplish the same number of business meetings: The average CES attendee has twelve meetings at the show, which means CES attendees save more than seven hundred million miles of flight in business trips that they otherwise would have to take.

They come to test the mettle of the people with whom they are doing business, to shake their hands, look in their eyes, and assess whether a company's product matches a company's hype. Perhaps most important, they arrive because relationships matter in business and, despite the worldwide reach of the Internet, a relationship cannot only be electronic. It must be personal.

This personal component to the International CES—or any trade show, for that matter—is what makes it a living, breathing entity. It's an experience that requires five senses. Some may scoff and wonder why in the age of technology and the Internet live, face-to-face events even exist. Yet they not only exist, they also prosper, because people, relationships, and firsthand impressions matter. Five-sense interaction beats the Internet for creating a big-picture view, allowing serendipitous discovery, developing trust, and enabling the evaluation of people and products.

Bill Gates once told me that Microsoft was made by trade shows like CES and COMDEX. Indeed, every large company started as a small business that needed to attract investors, partners, and customers quickly and efficiently. The best place to accomplish all three? At a trade show. Indeed, the entire CES is run with a basic principle: that anyone with an idea should be able to present it inexpensively to potential investors, buyers, partners, and media. Companies and careers have been made at trade shows—and they have changed our world as well.

Marketing executives at businesses across industry sectors understand the value of trade shows. In a recent study conducted by the nonprofit organization the Center for Exhibition Industry Research (CEIR) Foundation, 99 percent of surveyed executives said exhibiting provides unique value not offered by other marketing channels.[1] Business executives understand trade shows' importance, because exhibitions drive business. Another CEIR study conducted by Oxford Economics in 2010 found that each of the top exhibitions in the United States created, on average, $82 million in business-to-business sales among exhibitors and attendees.

The U.S. business-to-business exhibition industry is alive and well, and it offers companies a gateway to reaching their markets. The 2010 CEIR Exhibition Industry Census documents that there are approximately nine thousand business-to-business exhibitions in the United States. The CEIR Index estimates that close to 1.5 million companies exhibited in 2011 and roughly sixty million people attended.[2]

The power and purpose of trade shows in our rapidly connected world continue to grow with each year. Far from being a relic of a bygone age, today's trade show remains the premier event for learning about, interacting with, and maybe even striking a deal with the next generation of innovators.

But not every trade show survives. COMDEX is gone. After a period of decline, COMDEX was sold to a string of owners before eventually refashioning itself into just a Web presence. Why did this happen? COMDEX made mistakes—and CES had a killer strategy.

International CES: The Ninja Trade Show

WHEN I TOOK THE REINS OF CEA IN THE RECESSION OF 1990–91, the eventual demise of the summer show in Chicago, of our relationship with the video gamers, and of COMDEX were all in the future. I had a job to do, and that was keeping CES alive. But in the back of my mind a larger goal loomed: making CES the grandest and most popular consumer electronics show in the world. I knew we needed a strategy that reflected my ambitions. What I didn't know was that this strategy would encapsulate the very ninja-innovator qualities I have described in this book. And so I will attempt to reframe our strategy with those qualities in mind:

1. NINJAS FOLLOW A CODE OF CONDUCT: HONESTY.

This seems so simple, but it is vitally important in business and in life. Even if it's painful, you can never go wrong with honesty. Honesty is moral, ethical, and of course almost always the right thing to do. It is also easier. If you are honest, you don't have to remember who knows what and worry about the truth coming out.

One of our most painful steps toward honesty was shifting to accurately reporting how many people attend the International CES—a key method by which our customers measure us.

In the old days, we used to estimate our attendance by counting the number of show badges we printed ahead of time, and then we would assume that 20 percent of the people did not show up. Over the years, we had devised a formula: 80 percent of our pre-registration plus the people who registered at the venue was a good estimate of actual attendance. We would take that number, and rather than round off, we would make it an odd number so it would appear precise. We called this "actual estimated attendance." It wasn't dishonest, but as we knew quite well, it greatly exaggerated CES attendance.

This never seemed right to me. I didn't want to conduct a trade show that wasn't popular. But I felt it was more important to establish credibility over popularity. So to get a more accurate number we asked everyone who had preregistered to pick up a badge holder on site. This way we actually counted every person who came to our registration area and physically picked up a badge holder or registered on site. We also took the unusual move of establishing our credibility by hiring an outside independent auditor to verify that the people we said were there actually attended.

The following year, we saw a shockingly huge falloff in our reported attendance numbers. Instead of the 80 percent figure we were using, we discovered that only about half of those who preregistered for badges were really attending our show. This was quite a scary period. We didn't know what to do and debated how and whether we should release this data showing a decline in attendance from prior years' numbers. We knew the attendance hadn't fallen at all—just that our counting had become much better. It was an ethical dilemma. I weighed the option of sticking with a number I knew to be false against the certainty that publicly issuing an attendance number dramatically lower than prior years' attendance would make the show look like it was heading off a cliff. After much debate and thought, I decided we should be honest and transparent. We could always explain why there was a huge difference.

Of course, as soon as we released the much lower results, we took quite a negative hit in the press. Although we tried to explain that we changed how we counted people, the press reports were uniformly bad and devastating. Even the Las Vegas cabdrivers had heard the news reports and were saying that our show had shrunk dramatically in size. To us it was both frightening and humorous: frightening because the negative perception threatened the momentum of our show, but humorous since we knew that the show attendance from one year to the next had not really changed.

But by choosing honesty, we had the credibility to call for all trade shows to do the same. You could say it became a cause for me. I urged the entire trade show industry to shift to independent and honest reporting of attendance numbers. I got some traction and many shows now audit. If events are to compete against other media for marketing dollars, they need objective measurable audience data. Broadcasters have Arbitron and Nielsen, magazines and newspapers have the Audit Bureau of Circulations, and the Internet has comScore and Google Analytics to gauge page views. Why shouldn't events have audited attendance?

Yet, even today, a majority of U.S. shows do not audit. The International CES thus still has a strong advantage over our domestic and international competitors because we independently audit our show and our competitors do not. So if any of them claim huge attendance, take it with a large grain of salt, especially the Europeans! They count someone as a new and different person every day they attend an event.

Of course, honesty is both easy and wonderful when your numbers are increasing. But it is difficult when numbers are in decline. When we have to announce lower results it is a difficult pill to swallow. Nonetheless, I believe that honesty is the best policy. It allows you to deal consistently with good and bad situations. It builds trust, saves energy on deception, and is the ethical and right thing to do.

For us, this strategy was quickly rewarded. In 1998, when IBM, then one of the largest exhibitors in COMDEX, announced that it was dropping out, it specifically mentioned the fact that COMDEX would not independently audit its attendance numbers. Like I said, although it can be painful, honesty is never a bad policy.

2. NINJAS PAY IT FORWARD: TREAT OTHERS AS YOU WANT TO BE TREATED

Fortunately for CES, COMDEX ignored and mistreated its customers. At a panel session on which I participated with a former COMDEX president, years after his show collapsed, he admitted that they did not care about their customers. Rather, he said, COMDEX focused more heavily on financial issues, which included squeezing as much profit from the show as possible so they could eventually sell the show. No customers want to be treated poorly, and we strive to be decent and caring with our customers. More, most of the big companies who exhibit are CEA members, and we behave as if they are our bosses.

But over the years, we took being nice to a higher level. Of course, we were always polite, surveyed our customers extensively, and even strived to maintain personal relationships and connections. But we were never satisfied; we kept trying to step up our game. To that end, we undertake a tremendous amount of research. We ask our customers what they liked and didn't like and how we could do things better.

Second, we use a sales database, Salesforce.com, which allows us to track the history of all our sales contacts with each customer. We also encourage a little bit of competition among our sales employees. Every customer is fair game for any salesperson if they have not contacted them within thirty days; after that, other sales reps can contact them.

Third, we do not employ the outdated practice of "telling." As sales expert and bestselling author Jeffrey Gitomer says, "People do not like to be sold, but they love to buy."[3] If you are too busy *telling*, you only can hope that the customer is sold on your product. However, when you are effectively *selling*, you allow the customer to buy or to invest on their own terms.

Telling all too often involves insincerity. You enthusiastically flaunt the value of a product from *the seller's* perspective, which assumes that it makes perfect sense for its customers. That might be well and good, but it's only half of the picture. No one cares about how great your product is until they understand how great it is to *them*. The old salesman trick "Sell me this pen" really means "Tell me why I *need* this pen." Forget the pen; focus on the customer's needs. The same holds true for CES. We can wax poetically all day long about how great CES is, but until our prospects understand why it's great to *them*, they might as well be holding up a sign to us that says "So what?" We haven't earned the right to sell to anyone if we are only *telling* them what is great about us and our products.

Real selling involves asking questions to find out what matters most—ninjas are always gathering information—and then presenting your product as a series of *personalized* features and benefits relevant to your customers. This forms the basis for relationships—and the CES now has a lot of them. Indeed, I am proud to say that the CES continues to be ranked highly as an organization that focuses on our customers' needs, rather than assuming we know what's best for them.

Fourth, and perhaps most important, our sales vice president Dan Cole created a special customer-centric program that proactively focuses on real-time needs and desires to provide quick solutions for our clients. We call it SURE, and it stands for Sense of

Urgency, Responsiveness, and Empathy. We believe it's important to anticipate problems and demonstrate our sincere desire to solve them quickly. We want our customers (and potential customers) to know that we see their issues as if we were in their shoes. They expect us to respond quickly and we do. Our main goal is to mitigate angst and frustration before they happen. So we respond to our customers with a sense of urgency and empathy and it works well for us. Indeed, other corporations have adopted our SURE approach.

3. NINJAS TURN AN ADVERSARY'S STRENGTH AGAINST THEM

Although COMDEX was principally a computer show, it aggressively courted consumer electronics companies. Our response was to broadly define consumer electronics to include computers, IT, and anything having to do with the Internet. Our strategy was to go after COMDEX's strength, just as it was going after ours, rather than leave them the market.

But to beat COMDEX at its own game, we knew we had to do this carefully and deliberatively. Specifically, we defined the CES by the speakers who would deliver the keynote address. Led by our show vice president Karen Chupka, we had a major breakthrough when, in 1998, we persuaded Bill Gates to be a keynoter, joining Scott McNealy of Sun Microsystems. I will never forget seeing Gates walking alone into a rehearsal, his head buried in a magazine as he entered the hotel ballroom. Each year Gates's presentation and our investment in the stage set would grow, and we would attract more attendees and press and fill increasingly bigger rooms.

Most significantly, we used Gates's keynote to promote the CES as a major industry event—it wasn't every day that Bill Gates addressed the masses. This allowed us to attract other significant speakers, especially from the IT world, including Oracle's Larry Ellison and Cisco's John Chambers. We captured the IT space, and as new IT products like smart phones, laptops, tablets, and computers were introduced through consumer channels, we were attractive to companies seeking to make their mark with the press, power retailers, and the financial community. The COMDEX formula, directed at the geek and CIO crowd, faded in importance, and we stole their market share.

The lesson for me in this is that although only about 2 percent of our attendees (around three thousand people) actually attend the keynotes, the positioning and marketing of our event revolves heavily around the keynotes. For this reason even today we are very careful about whom we select as keynoters because we want to use them to define the event and reinforce the importance of the show. This strategy has had a great side effect: Major-company CEOs *want* to speak at the CES; it is quite the career marker for them and their companies because of the global coverage. We also use the keynotes to broaden the definition of consumer electronics to include wireless, video games, automobiles, and even the entertainment community. I consider Karen's vision of the enhanced nature and position of the keynotes to be one of the major reasons the International CES is the most successful technology event in the world.

4. Ninjas Can Reinvent Themselves Nimbly
With one simple word, we changed ourselves.

After the end of our summer show in Chicago, we simply called our winter show in Las Vegas the Consumer Electronics Show, or CES. But then one of our marketing professionals suggested we add one word to our name: *international.* I recall hearing the idea and approving it but not thinking it was that big a deal. It turned out to be among the most significant changes we made, and one of the easiest. When you are international, you are vastly more important. Simply by adding the word *international* in front of CES, we fundamentally transformed the nature and perception of our event. With that one-word addition and a little marketing, we started attracting additional international visitors to our show—which now number more than thirty thousand each year. This made the show more important to our exhibitors because they could greatly expand their sales by attracting overseas buyers.

Since then, we have worked hard to embrace and expand international attendance by traveling the world and persuading more customers. By targeting overseas press and catering to foreign needs—like hiring more interpreters—we lived up to our clever marketing trick: CES *is* international. More, we have worked with the Las Vegas Convention and Visitors Authority and the U.S. Commerce Department to help us seek and entice foreign buyers and important media to visit Las Vegas.

Our long-term investment in attracting international attendance has paid off handsomely. The 2012 International CES attracted more than thirty-five thousand people from outside the United States to Las Vegas. This huge international contingent bolsters the national economy, increases the importance of the event, and enables our global exhibitors to use international marketing dollars to support their presence.

5. Ninjas Take Advantage of Their Surroundings
One lesson from tae kwon do was to always be aware of your surroundings. Look around wherever you are and observe. What are the threats? What are the opportunities? What is the escape route?

And, especially relevant to this discussion, what potential weapons are lying nearby in the event of an attack? A ninja is never without options, because he can use everything to his advantage.

What a trade show does is sell exhibiting companies the ability to meet new and existing customers in a cost-efficient manner. But in my first days as CEA head, the customers were evaporating. Our traditional attendee audience of thousands of small independent "mom and pop" retailers was going out of business as more consumers favored "big box" stores like Best Buy and discount stores such as Costco. More troublesome, Internet retailers, including Amazon, Crutchfield, and Newegg, were also growing and attracting customers. Instead of thousands of retailers, we had just a handful of huge buyers. We used to joke that 90 percent of the buying power in our industry could fit into a small room.

This evolution of retail from Main Street to the strip mall was beyond our capacity to change. The free market was doing what it does, and rather than blow against the wind, we decided to adapt. So we developed a strategy that focused on other essential audiences for our exhibitors: the press, the investment community, and related industries and potential partners. We developed tactics to attract each audience that included targeted marketing, relevant conference programming, and various VIP designations.

The consolidation of the buyer community is even more pronounced than it was when we first noticed it. And yet, the 2012 International CES was our most popular show to date.

6. Ninjas Are Always Seeking Allies—Even Among Their Enemies

Certainly no man is an island, and in the trade show world, partnering is everything. The book *Co-Opetition: A Revolutionary*

Mindset That Combines Competition and Cooperation; the Game Theory Strategy That's Changing the Game of Business, authors Adam Brandenburger and Barry Nalebuff describe the advantage of cooperating with your competitors. The old adage "Keep your friends close and your enemies closer" is another way of describing this vital strategy.

The International CES succeeds because it is a huge tent. We embrace everyone connected with our industry. We attract more than a hundred related associations to join us as "allied associations." We give them recognition and special treatment in return for their support. We battle many of them in Washington, yet we embrace them in Las Vegas. More, we are always civil and we support them and their events when and how we can. And if we are having a panel or conference on areas in which we disagree, we invite them to participate so they can share their viewpoint.

We also expand our show's breadth by partnering with websites, publishers, experts, and other groups. We have such a broad innovation-based event that we attract so many disparate industries. We do not and cannot possess expertise in all these areas, so we reach out to those who are knowledgeable, share our values, and create partnerships. Our theory of partnering is that both parties must win. So we generally try to structure our partnerships so we are not micromanaging each other's expenses and both have incentives to grow top-line revenue.

I love these deals because I get to meet some of the most entrepreneurial and innovative people in the world. They bring energy, excitement, and novelty to the International CES. Thus, we created our Living in Digital Times section with entrepreneur Robin Raskin, our government-focused event with procurement expert and close friend Don Upson, and our Digital Hollywood program

with the colorful Victor Harwood. For each CES, we have between twenty and thirty active partnerships that extend our brand
and provide us with partners whose only interest is to make the
International CES better.

7. Ninjas Have Energy and Passion

We always keep the International CES fresh and exciting.

No one likes to be bored. And certainly no one wants to go
to a boring event. Even in business, whether someone will invest
their time and money to attend and participate in any endeavor is a
matter of discretion. I know flying is not the fun it used to be, and
while Las Vegas is fantastic for many, the plane trip, the expense,
and the time away from home and the office make it costly to participate. So a big part of our strategy is to give everyone multiple
reasons to justify their attendance.

I start with the assumption that the decision is close to 50-50 as
to whether someone will attend. So we work very hard to make everyone feel welcome and comfortable. We also assume that many
people are motivated also by fear and greed. People are fearful they
will miss something important, and they're greedy in the sense that
they want to see new opportunities to make money.

So before the show, we use marketing, the media, and special invitations to ensure that every qualified potential attendee
receives multiple reasons from several sources to attend the show.
This includes giving our exhibitors the opportunity to reach these
attendees, building up excitement for the show, spacing out press
announcements about major and exciting developments, and ensuring a buzz is always present in the months leading up to the
show.

More, as the show producer, we can't just rely on our exhibitors to be alluring—we have to add some sizzle to the steak! I have a rule that every year we must add at least three new compelling and exciting special events, areas, or happenings to CES. We usually have many more, but we often highlight and promote only a few. We not only have to get people to register for our event—we want them to get on a plane and show up!

8. NINJAS NEVER SETTLE

Even though we succeeded in making the International CES the dominant and most popular trade show in our industry, we aren't content. We work hard to ensure a great experience for all who come.

Getting people to attend our event is just the start. We also want them to have a great experience while they're there so they will come back next year. We do this by surveying our attendees in depth and seeing what we can improve on. One of the most important questions in the survey is a rating we ask our guests to give the CES staff. We always await the results and want to see that the scores are rising year over year.

To ensure this, we have training sessions for staff before each event on every aspect of the show, and we emphasize how each employee is an ambassador. We even bring on many of our former employees to work the show because they know our events and our customers, and because they have the passion and the right attitude. We even go farther and have a special program for employee family members we call Share the Love, in which they can help out the show, get paid, and provide information and other services to guests.

Every year, before each show we extensively train and even test all staff attending the show. At "graduation," I give a pep talk. I talk about how each and every CES worker is part owner in the show—they share our success, they share our failure. The International CES is a means to accomplish our vision of improving the lives of people through innovation. I communicate what is at stake for our customers—especially the smaller ones—who risk so much to be at the event. I describe the 50-50 view of how people make the decision to attend events and that every touch or contact they have with an attendee will create an impression. I tell them *they* determine whether our event succeeds or fails.

The actual show is just the culmination of years of planning. Before each CES, we consider every aspect of the event, from the buses and their frequency and routes, to the thickness of the carpeting, to the layout of the floor plan and the website for registration and information, to meetings with exhibitors. We aren't perfect, but we are always improving, and as the premier innovation event in the world we work hard to keep the event a phenomenal and worthwhile experience for everyone who comes.

The Thrill of Victory

When I was given the reins of CEA and its premier event, the International CES, I took on an enormous task. I was presented with the dual challenge of keeping CES relevant as well as fending off our biggest competitor, COMDEX. I turned that challenge onto my team: Let's not just stay relevant; let's beat COMDEX.

I'm proud to say that they delivered.

And because of that I was able to learn about how to build a successful organization. We made mistakes—nearly fatal mistakes. But we grew from those errors. More important, we learned from the missteps of our competitor. It was this process that taught me the skills of the ninja innovator. It taught me how to succeed.

If I'm able to pass judgment on the successes and failures of other enterprises and individuals today, it's because I've not only learned from my members what success requires, I've experienced it first-hand. At the time, I did not know that these skills were particularly reflective of the ancient ninja warrior. Now I understand that what made the ninja such a successful entity makes us all successful.

The Heart of the Ninja

BY THE EARLY EIGHTEENTH CENTURY, REFERENCES TO NINJAS IN the historical record all but cease. No one is quite sure what happened, but the likely explanation is that as Japan became a more unified country with fewer squabbling fiefdoms, the ninja's importance waned. By the nineteenth century, the ninja had entered the realm of legend. Wild stories about their magical skills—walking on water, being invisible—proliferated and it was difficult to separate the historical ninja from the mythical ninja. Once Hollywood got ahold of them in the twentieth century, the ninja became an icon: one of the greatest warriors of a faraway land. Real-life ninjas, such as Hattori Hanzo and Fuma Kotaro, became popular fictional characters whose abilities in espionage and with the sword captured audiences' imaginations. And that's pretty much where the "shadow warrior" remains today.

I'll be the first to admit that applying the adjective *ninja* to a process is not particularly original. To be a "ninja" in anything is to be the best in class. Unfortunately, the label is so widely used that it's lost much of its historical resonance. Everyone knows what a ninja is, but few can actually tell you who and what they were. Part of my purpose throughout this book has been to illuminate the historical ninja, because I firmly believe that what set the ninja apart in feudal Japan also differentiates the successful innovator or enterprise in twenty-first-century America. That, and also because the historical, fact-based ninja was just downright cool.

But this book was never intended to be a historical study, and I must beg forgiveness from any scholars who recognize the errors that may have crept into my writing. As I said at the start, this is a book about success and how to achieve it. But I hope that my appreciation for failure came through just as strong. For all of the case studies I have mentioned whose example I positioned as what not to do, it was not my intention to single out any individual or enterprise as an object of ridicule. Rather, my hope is that these enterprises learned from their failure and became stronger and better. After all, as I'm sure the historical and the mythical ninja warrior would tell us, it is through failure that we learn.

In both my career and my study of tae kwon do, I learned that success does not happen overnight. Like a novelist working on a book, it takes a long time and lots of hard work to succeed. At CEA, I deal with the federal government almost every single day. Trust me, this requires some patience. But I look back on my experiences, notably the HDTV and International CES experiences, and I'm comfortable with the knowledge that most things worth achieving happen over a long—sometimes painfully long—period. I also look at my family. Nothing trains you for patience quite like caring for a newborn, a toddler, an adolescent, and a teenager. A parent's work is never done. Why do it? Because there are those moments when, looking on your achievement, you discover true happiness. All the pain, all the effort, all the late, late nights, suddenly evaporate in the moment when you gaze upon your child and experience true contentment.

Even though I am a black belt, I am quite conscious of my own limitations. But that's not the point. Just because I can't do everything doesn't mean I can't do many things well. Spending hours, years, training and studying for the black belt taught me that success—as an individual or an enterprise—is the culmination of years of hard work. It takes training, discipline, and failure to achieve your potential. But I learned. To be a ninja requires a dedication to yourself, a respect for the idea that you can succeed. It's not easy, but nothing worth acquiring is easy.

By now you know that ninjas are those who change the world. Ninja innovators think about what can be, rather than what is. They challenge themselves to make things happen. They live life fully. They make goals. They achieve. They stretch themselves, and in doing so attain great joy.

Anyone can be a ninja. Being a ninja is a way of looking at and approaching the world. It is a philosophy of being your best. A way of solving problems. A commitment to making yourself and your enterprise better.

Every person, group, business, and government can develop a ninja culture. Any person or entity can embrace a strategy of innovative thinking and breaking barriers to find success.

Ninjas imagine the future and then create it. My goal in this book was to inspire you to create your future using ninja techniques. I hope you create your future and share with me your results.

ACKNOWLEDGMENTS

My name is on the cover and I am responsible for any mistakes, but I thank those whose guidance, decisions, and efforts turned this dream into a reality.

It took a ninja strike force to produce this book, and every role was important and contributed to our mission's success.

This book began when my four grandparents escaped horrible conditions. They were each ninjas—confronting harsh realities and acting decisively, adapting, and, though not prospering financially, certainly surviving and giving their kids a better life.

My mother's parents, Max and Manie, left Russia in a hay wagon and got married only after settling in Montreal. My dad's mother, Jane, escaped the anti-Semitism of Poland and emigrated to New York, where she met Romanian Leon, my grandfather, and they launched a tiny grocery store, the Broadway Dairy, on Broadway and Ninetieth Street in Manhattan.

One summer, while serving as counselors at an upstate New York summer camp, my dad, Jerome, became enchanted with Mildred's laugh, and they married the following January. Dad had returned from service in World War II and received a graduate degree in education under the GI Bill. Mom and Dad were teachers but held extra jobs to expose us to music, theater, sports, the ocean, and 4-H. After almost fifty years of marriage, Dad took care of my mom, who was suffering from Alzheimer's, for a dozen years in our family home until she died. Dad passed away a year later.

Dad was my first mentor. He spent thousands of hours with me—from weekly trips to the library to thousands of rounds of chess, cribbage, and other card games. He insisted that I learn the math behind cards. He never denied my requests for time and advice nor did he judge or direct me. He used questions to ensure that I thought through options and consequences.

Dad led with wisdom, but Mom grilled me on current events and etiquette—and had me walk for hours with a book on my head to improve my posture. She was tough but she taught me to respect my body, to exercise, to eat healthy, and to balance hard work with time at the beach. She composted and avoided meat before it was trendy and taught me to be a lifelong learner.

ACKNOWLEDGMENTS

My three brothers, Eric, Ken, and Howie, and I grew up with an ethos of hard work, frugality, love for family, and value in education. Mom and Dad also gifted us with self-confidence and a sense that we could solve any problem. They gave us a ninja upbringing of creative problem-solving.

I had many great teachers and mentors who changed my life. In high school, my sociology teacher, Mr. Dinapoli, challenged me to reject status-quo arguments. At Georgetown Law School, my contracts professor, Richard Gordon, impressed on me how chance and action determine the course of life. Congressman Mickey Edwards gave me a Capitol Hill job, where I learned how government works and how Americans all have different views.

Former Federal Trade Commissioner Jim Nicholson gave me a clerkship at his law firm and taught me to quickly analyze merger prospects for our Wall Street clients. I learned to get competitive info, draw conclusions with partial data, and analyze judges as humans—all pre-Internet. Postmaster General, under President Kennedy, J. Edward Day took me under his wing at the law firm Squire Sanders and let me manage a big client, who soon offered me a job and put me in charge, the position I still hold today.

I began at the Consumer Electronics Association (CEA) under the tutelage of World War II Purple Heart recipient Jack Wayman. He thinks out of the box and is one fast-moving ninja. He not only launched the International CES but also approached COMDEX owner Sheldon Adelson about a deal to build a Las Vegas building, and then he turned the project over to me. Dealing with the legendary Adelson was baptism by fire, but I learned then that visionary ninjas do what it takes to realize their dreams.

I also learned much from association leader Pete McCloskey who elevated me to become his general counsel. Pete taught me that consulting colleagues and listening on big issues are how to get things done, that nastiness is not helpful, and that family should come first.

I had numerous directors from the industry who helped transform me from lawyer to leader. Jerry Kalov taught me about life and encouraged me to take risks—promising his support when I failed . . . which I often did. Jack Pluckhan showed me Japanese business culture and the importance of patience, respect, and loyalty. Peter Lesser served as a tough sounding board and taught me that New Yorkers are more direct than other Americans.

CEA is a ninja organization because we have strong leaders and a culture of street smarts, opportunism, ethics, and support for the newest entrants in an industry that continually redefines itself. Among them are ninja legends in their own right. I met Darrell Issa when he had a union problem as a CES exhibitor. I resolved the problem. Darrell engaged with us, moved up our leadership chain, and led our effort to break from our parent organization. Today, Darrell is a tough and effective congressional committee chairman.

Joe Clayton headed RCA, Global Crossing, Sirius, and now Dish. He is strong and smart, and a brilliant marketer. Under his leadership, the association grew, changed its name, and began the quest for relevance both in the industry and in Washington.

Kathy Gornik, Loyd Ivey, and John Shalam have been amazing by so many measures. They all are living the American Dream, running great companies that they began at CES. They are passionate about CEA and CES, and they carry the torch for ensuring that any entrepreneur with an idea can exhibit at CES inexpensively as well as meet buyers, media, and investors.

Randy Fry has been a fabulous chairman who is always pushing the envelope and encouraging the CEA's efforts to advocate for innovation. Along with past chairs, Pat Lavelle and Gary Yacoubian, we have had stellar leadership and support for advancing innovation as a strategy for our industry and our nation. Of course, *Ninja Innovation* would not have been published without their support and that of their colleagues on the 2012 CEA Executive Board: Jim Bazet, Denise Gibson, Robert Fields, Jay McLellan, Phil Molyneux, Daniel Pidgeon, George Stepancich, Steve Tiffen, and Mike Vitelli. Indeed, CEA owns the book *Ninja Innovation*!

I must acknowledge the teachers at Kim's Karate in Annandale, Virginia. They taught me not only technique but also discipline. I continue my physical training today under the tutelage of Jeff Strahan and Wolf Gottschalk of Fitness Image Results. A shout-out to my fellow CEA Boot Campers—a daily exercise program at our office! Your encouragement and support have been inspiring.

I also want to thank a few groups to which I give my time. No Labels is trying to change American politics and put the nation before political parties. Go to www.nolabels.com and join right now!

The World Electronics Forum gathers my colleagues from around the world. We are pushing worldwide innovation, and it matters!

The Northern Virginia Tech Council, led by the amazing Bobbie Kilberg, like its counterparts around the country, fights for innovation as a cause. I have visited and spoken before many of these groups during the past two years, and I am thrilled with the passion and ideas of these innovators. Our future's bright if we unleash their power.

Similarly, Washington has brilliant people leading tech organizations that are making a difference in our innovation future: Grant Seiffert of the Telecommunications Industry Association, Bruce Mehlman of the Technology CEO Council, Robert Holleyman of BSA/The Software Alliance, Ed Black of the Computer & Communications Industry Association (CCIA), Dean Garfield of the Information Technology Industry Council (ITIC), and Jay Timmons of the National Association of Manufacturers.

Some thoughtful think-tank and public-interest groups help push good innovation policy: Public Knowledge (led by the indomitable Gigi Sohn), the Cato Institute, the Heritage Foundation, and the Electronic Freedom Foundation.

ACKNOWLEDGMENTS

A book requires many hands and experts. Inside CEA, Laurie Ann Phillips provided tons of heavy lifting and made sure the process worked. Jeff Joseph and Laura Hubbard helped in the op-eds and book process. Julie Kearney, Karen Chupka, Michael Petricone, Dan Cole, and Brian Markwalter read and commented on various sections. Susan Littleton, Michael Brown, and John Lindsey provided huge marketing and design support, and advice. My amazing and wonderful assistant Jackie Black made sure that the process worked every step of the way. And our sharp-shooter lawyers John Kelly and Kara Maser, along with our phenomenal COO, Glenda MacMullin, supervised the business arrangements. Glenda also managed the CEA ship so I could write this book.

Outside commenters on intellectual property included musician lawyer Bob Schwartz and business copyright expert David Leibowitz. I received wonderful input on the digital television section from HDTV legend Peter Fannon. Bob Schwartz and Seth Greenstein from Constantine Cannon weighed in on the HD radio section.

An incredible thank you is owed to editor Blake D. Dvorak of the Pinkston Group. He made this happen. His organization, research on ninjas, and rewriting of my submissions have made this book more readable and interesting. His colleagues Christian Pinkston and David Fouse also helped make this dream become a reality.

Of course, my publisher, William Morrow/HarperCollins, has been a delight. Creative, professional, transparent, and willing to negotiate. Thank you especially to Executive Editor Peter Hubbard, as well as to Andy Dodds, Cole Hager, and Tavia Kowalchuck. Also a big thank you to Lynn Grady, whose creativity in marketing and enthusiasm made the entire project so enjoyable.

And last but most important, I want to thank my family. My wife, Dr. Susan Malinowski, never hesitated to support my efforts to write this book, despite her own schedule of performing groundbreaking research, handling a busy medical practice, obtaining a patent, building our house, raising our four year old, being pregnant and then giving birth, and running a 10K race all in this very busy past year. But none of this could have happened, including quiet time for my writing, without the daily lengthy and generous childcare help from her parents, my wonderful in-laws, Drs. Jola and Edward Malinowski.

I am blessed with the love of family, the help of friends, and the support of colleagues and volunteers at CEA. They have been great to me and key in the success of an amazing industry. We are bound with a shared passion that innovation is the strategy that will make us all better. I hope you will join CEA's Innovation Movement at www.declareinnovation.com.

NOTES

INTRODUCTION: THE WAY OF THE NINJA

1. David Progue, "Sampling the Future of Gadgetry," *New York Times*, January 11, 2012, http://www.nytimes.com/2012/01/12/technology/personaltech/in-las-vegas-its-the-future-of-high-tech-state-of-the-art.html?pagewanted=1&_r=1&ref=internationalconsumerelectronics showces.
2. William Scott Wilson, *Ideals of the Samurai: Writings of Japanese Warriors* (Valencia, CA: Black Belt Communications, 1982), 47.
3. Stephen Turnbull, *Ninja: AD 1460–1650* (Oxford, UK: Osprey Publishing, 2003), 17.
4. Clayton M. Christensen, *The Innovator's Dilemma: The Revolutionary Book That Will Change the Way You Do Business* (New York: Harper-Business, 2011).

CHAPTER 1: YOUR GOAL IS VICTORY

1. Fortune 500 "Top Companies" list, CNN, http://money.cnn.com/magazines/fortune/fortune500/2012/performers/companies/biggest/.
2. Fortune 500 list, "19. International Business Machines," CNN, http://money.cnn.com/magazines/fortune/fortune500/2012/snapshots/225.html.
3. Verne G. Kopytoff and Laurie J. Flynn, "PC Makers Are Seeing a Slowdown," *New York Times*, May 17, 2011, http://www.nytimes.com/2011/05/18/technology/18compute.html.
4. Darryl K. Taft, "IT and Network Infrastructure: IBM at 100: 20 Technologies that Soared and 10 That Failed," eWeek.com, http://www.eweek.com/c/a/IT-Infrastructure/IBM-at-100-20-Technologies-That-Soared-and-10-That-Failed-348925/.
5. Thomas L. Friedman, "Our One-Party Democracy," *New York Times*, September 8, 2009, http://www.nytimes.com/2009/09/09/opinion/09friedman.html.

NOTES

6. "China to Overtake Silicon Valley, Claims Report," *China Real Time Report* (blog), *Wall Street Journal*, June 27, 2012, http://blogs.wsj.com/chinarealtime/2012/06/27/china-to-over-take-silicon-valley-claims-report/.

7. James McGregor, "China's Drive for 'Indigenous Innovation,'" APCO Worldwide, http://www.uschamber.com/sites/default/files/reports/100728chinareport_0.pdf.

8. Xinhua News, "Nation's 10-Year Education Plan Unveiled," *People's Daily* (English edition), July 30, 2010, http://english.people.com.cn/90001/90776/90785/7086363.html.

9. Hisense press release, "Creating a Dialogue: Hisense Collaborates with MIT," January 9, 2012, http://www.hisense.com.my/news.php?content_id=34.

10. Shamim Adam, "China May Surpass U.S. by 2020 in 'Super Cycle,'" Standard Chartered Says," Bloomberg, November 14, 2010, http://www.bloomberg.com/news/2010-11-15/china-may-surpass-u-s-by-2020-in-super-cycle-standard-chartered-says.html.

CHAPTER 2: YOUR STRIKE FORCE

1. Eric D. Isaacs, "Forget About the Mythical Lone Inventor in the Garage," *Slate*, May 18, 2012, http://www.slate.com/articles/technology/future_tense/2012/05/argonne_national_lab_director_on_the_myth_of_the_lone_inventor_in_the_garage.html.

2. Adam Cohen, "Going, Going, Gone: Meg Whitman Leaves eBay," *New York Times*, January 25, 2008, http://theboard.blogs.nytimes.com/2008/01/25/going-going-gone-meg-whitman-leaves-ebay/.

3. Meg Whitman Alumni Achievement Awards profile, Harvard Business School, http://www.alumni.hbs.edu/awards/2008/whitman.html.

4. Timothy Sexton, "eBay's Meg Whitman: America's Greatest CEO?" Yahoo!, February 20, 2007, http://voices.yahoo.com/ebays-meg-whitman-americas-greatest-ceo-209493.html?cat=3.

5. Gary Dessler and Jean Phillips, *Managing Now* (New York: Houghton Mifflin, 2008).

6. "Meet eBay's Auctioneer-in-Chief," *Bloomberg Businessweek*, May 28, 2003, http://www.businessweek.com/stories/2003-05-28/meet-ebays-auctioneer-in-chief.

7. Dawn Kawamoto and Corey Grice, "eBay Roars into Public Trading," CNET, September 24, 1998, http://news.cnet.com/eBay-roars-into-public-trading/2100-1001_3-215908.html.

CHAPTER 3: IN WAR, RISK IS UNAVOIDABLE

1. Henry Samuel, "France's Académie Française Battles to Protect Language from English," *Daily Telegraph* (UK), October 11, 2011, http://www.telegraph.co.uk/news/worldnews/europe/france/8820304/

230

Frances-Academie-francaise-battles-to-protect-language-from-English
.html.

2. "Nuclear Power in France," World Nuclear Association, http://www
.world-nuclear.org/info/inf40.html.

3. "History," Thales, http://www.thalesgroup.com/Group/About_us/
History/.

4. Christine Winter, "GE Bids Adieu to Electronics, RCA," *Chicago Tribune*,
July 23, 1987, http://articles.chicagotribune.com/1987-07-23/business/
8702230825_1_rca-thomson-sa-zenith-chairman-jerry-pearlman.

5. Robert E. Calem, "Technology; Back from the Brink, RCA Is Forg-
ing a Digital Future," *New York Times*, October 23, 1994, http://www
.nytimes.com/1994/10/23/business/technology-back-from-the-brink-
rca-is-forging-a-digital-future.html?pagewanted=all&src=pm.

6. Peter J. Williamson and Anand P. Raman, "The Globe: How
China Reset Its Global Acquisition Agenda," *Harvard Business
Review*, May 10, 2011, reprinted at http://viewswire.eiu.com/index
.asp?layout=ebArticleVW3&article_id=1358056520&channel_
id=778114477&category_id=1138152913&refm=vwCat&page_
title=Article.

7. http://www.eetimes.com/electronics-news/4067823/Thomson-to-sell-
RCA-in-50-million-deal.

8. Nicolas Sarkozy, "Opening of the eG8 Forum: Address by Nicolas Sarkozy, President of the French Republic," G8 Information Centre, May
24, 2011, http://www.g8.utoronto.ca/summit/2011deauville/eg8/eg8-
sarkozy-en.html.

9. Cited in "The Implications of Optimism," NYU Stern School of Business, May 2004, http://pages.stern.nyu.edu/~alandier/pdfs/optimism
.html. Full report available at http://www.cepr.org/meets/wkcn/6/6620/
papers/thesmar.pdf.

10. Ibid.

CHAPTER 4: PREPARE FOR BATTLE

1. Yardena Arar, "Gates Wows COMDEX Crowd with Tablet PC," *PC-
World*, November 13, 2000, http://www.pcworld.com/article/34751-2/
gates_wows_comdex_crowd_with_tablet_pc.html.

2. Ephraim Schwartz, "Microsoft to Unveil Tablet PC," *PCWorld*, November 8, 2008, http://www.pcworld.com/article/34329/microsoft_to_
unveil_tablet_pc.html.

3. John Fontana, "Microsoft's History with the Tablet PC," *PCWorld*,
January 17, 2010, http://www.pcworld.com/article/187062/microsofts_
history_with_the_tablet_pc.html.

4. Jonah Lehrer, "Don't!: The Secret of Self-Control," *New Yorker*, May
18, 2009, http://www.newyorker.com/reporting/2009/05/18/090518fa_
fact_lehrer.

5. Stephen Turnbull, *Ninja: AD 1460–1650* (Oxford, UK: Osprey Publishing, 2003), 27.
6. "Profiles on Legal Permanent Residents: 2011," U.S. Department of Homeland Security, http://www.dhs.gov/files/statistics/data/DSLPR11c.shtm.
7. "Persons Obtaining Legal U.S. Permanent Resident Status by Region and Selected Country of Last Residence: 1820–2010," U.S. Department of Homeland Security, http://www.dhs.gov/xlibrary/assets/statistics/maps/lpr/lpr_map_icolr_1820_2010.pdf.
8. Vivek Wadhwa, "The Faces of Success, Part I: How the Indians Conquered Silicon Valley," *Inc.*, January 13, 2012, http://www.inc.com/vivek-wadhwa/how-the-indians-succeeded-in-silicon-valley.html.
9. Vivek Wadhwa, "Why Silicon Valley Should Fear India," *Washington Post*, November 22, 2011, http://www.washingtonpost.com/national/on-innovations/why-india-should-scare-silicon-valley/2011/09/14/gIQALjiolN_story.html.
10. "India Will Not Pursue Protectionism Policy: Telecom Minister Kapil Sibal to Foreign Electronics Manufacturers," *Economic Times*, February 28, 2012, http://articles.economictimes.indiatimes.com/2012-02-28/news/31108139_1_import-bill-preference-in-government-procurement-shyam-telecom.

CHAPTER 5: THE ART OF WAR

1. Bari Weiss, "Groupon's $6 Billion Gambler," *Wall Street Journal*, December 20, 2010, http://online.wsj.com/article_email/SB10001424052748704828104576021481410635432-lMyQjAxMTAwMDEwODExNDgyWj.html.
2. James B. Stewart, "When the Network Effect Goes into Reverse," *New York Times*, August 17, 2012, http://www.nytimes.com/2012/08/18/business/Sites-Like-Groupon-and-Facebook-Disappoint-Investors.html?_r=1.
3. Peter F. Drucker, *Management: Tasks, Responsibilities, Practices* (New York: Harper & Row, 1973, 1974), 62.
4. Josh Spiro, "The Great Leaders Series: Jeff Bezos, Founder of Amazon.com," *Inc.*, October 23, 2009, http://www.inc.com/30years/articles/jeff-bezos.html.
5. Daniel Lyons, "'We Start with the Customer and Work Backward': Jeff Bezos on Amazon's Success," *Slate*, December 24, 2009, http://www.slate.com/articles/news_and_politics/newsmakers/2009/12/we_start_with_the_customer_and_we_work_backward.html.
6. Associated Press, "Amazon: E-Commerce Success Story," CBS News, February 11, 2009, http://www.cbsnews.com/2100-205_162-706351.html?tag=contentMain;contentBody.
7. Bureau of Labor Statistics, "Local Area Unemployment Statistics: Un-

employment Rates for States," U.S. Department of Labor, http://www
.bls.gov.

8. ZDNet Research for IT Facts, "50% of Internet Traffic Passes Through
Virginia," December 11, 2003, http://www.zdnet.com/blog/itfacts/50-
of-internet-traffic-passes-through-virginia/5057.

9. Matt Scherer, "Shapiro Interview Pt 2," *AustinStartup*, April 4, 2012,
http://austinstartup.com/2012/04/shapiro-interview-pt-2/.

CHAPTER 6: THE NINJA CODE

1. Jerry B. Harvey, "The Abilene paradox: the management of agreement,"
Organizational Dynamics 3 (Winter 1974): 63–80.

2. Doug Gross, "Did Steve Jobs Kill Adobe Flash?," CNN, November 9,
2011, http://articles.cnn.com/2011-11-09/tech/tech_mobile_flash-steve-
jobs_1_html5-battery-hog-photoshop?_s=PM:TECH.

3. Andrew McLaughlin, "Google Goes to Washington," *Google Official Blog*, http://googleblog.blogspot.com/2005/10/google-goes-to-
washington.html.

4. Alex Fitzpatrick, "Google Spends More on Lobbying than Apple, Facebook, Microsoft Combined," *Mashable*, April 24, 2012, http://mashable
.com/2012/04/24/google-record-lobbying/.

5. Henry Juszkiewicz, "Gibson's Fight Against Criminalizing Capitalism,"
Wall Street Journal, July 19, 2012, http://online.wsj.com/article/SB1000
1424052702303830204577448351409946024.html.

6. Rob Bluey, "Months After Federal Raid, Gibson Guitar Still Faces No
Charges," *The Foundry* (blog), February 23, 2012, http://blog.heritage
.org/2012/02/23/video-months-after-federal-raid-gibson-guitar-still-
faces-no-charges/.

7. "Gibson CEO to Obama: Show Some Concern for Nation's Job
Creators," Fox News, September 8, 2011, http://www.foxnews.com/
politics/2011/09/08/gibson-ceo-to-obama-show-some-concern-for-
nations-job-creators/.

8. http://www.gibson.com/absolutenm/templates/FeatureTemplate
PressRelease.aspx?articleid=1359&zoneid=6.

9. James Oliphant, "Labor Board Drops Complaint Against Boeing over
S.C. Plant," *Los Angeles Times*, December 9, 2011, http://articles.la
times.com/2011/dec/09/news/la-pn-nlrb-boeing-20111209.

CHAPTER 7: NINJAS BREAK THE RULES

1. Scott Shane, "California Rules the Venture Capital Ecosystem," *Small
Business Trends*, December 22, 2010, http://smallbiztrends.com/2010/12/
california-rules-venture-capital-ecosystem.html; Marc Lifsher, "California Is by Far the Leading State in Luring Venture Capital," *Los
Angeles Times*, April 5, 2012, http://articles.latimes.com/2012/apr/05/
business/la-fi-mo-california-far-away-leader-20120405.

2. Jan Norman, "California Ranks 50th in New Business Creation," *Orange County Register*, August 28, 2011, http://jan.ocregister.com/2011/08/28/california-ranks-50th-in-new-business-creation/63331/.

3. Jan Norman, "Report: 254 Companies Left California in 2011," *Orange County Register*, March 2, 2012, http://www.ocregister.com/articles/moved-342887-companies-texas.html.

4. G. Scott Thomas, "Texas, North Dakota Are Rare States with Private-Sector Employment Gains," *Business Journals*, March 26, 2012, http://www.bizjournals.com/bizjournals/on-numbers/scott-thomas/2012/03/texas-north-dakota-are-rare-states.html?appSession=84998451895589&RecordID=&PageID=2&PrevPageID=2&cpipage=1&CPIsortType=desc&CPIorderby=Percent_change&cbCurrentPageSize=.

5. Edmund G. Brown, Jr., Governor of California, et al. v. Entertainment Merchants Association et al., 131 S. Ct. 2729 (2011), http://scholar.google.com/scholar_case?case=12960598670321445636&q=Brown+v.+Entertainment+Merchants+Association&hl=en&as_sdt=2,47&as_vis=1.

6. Marcia Fritz, "Put Pension Reform in California to the Vote," *Los Angeles Times*, June 26, 2012, http://www.latimes.com/news/opinion/commentary/la-oe-fritz-pension-reform-california-20120626,0,1584148.story.

7. Joseph M. Hayes, "California's Changing Prison Population," Public Policy Institute of California, April 2012, http://www.ppic.org/main/publication_show.asp?i=702; Chris Megerian, "Officials Announce Sweeping Overhaul of California Prisons," *Los Angeles Times*, April 24, 2012, http://articles.latimes.com/2012/apr/24/local/la-me-prisons-20120424.

8. "Generating Higher Value at IBM," IBM, http://www.ibm.com/annualreport/2011/ghv/#two.

CHAPTER 8: INNOVATE OR DIE

1. "Amazon: eBook Sales Surpass Printed Books," *Huffington Post*, May 19, 2011, http://www.huffingtonpost.com/2011/05/19/amazon-ebook-sales-surpas_n_864387.html.

2. Ronald Glover, "Hollywood Hopes Rise as Blu-ray, Digital Offset DVD Decline," Reuters, April 29, 2012, http://www.reuters.com/article/2012/04/29/entertainment-us-deg-idUSBRE83S0B120120429.

3. Philip Elmer-DeWitt, "Analysts: iPod Sales Expected to Decline 7.2% Year over Year," CNN, July 15, 2011, http://tech.fortune.cnn.com/2011/07/15/analysts-ipods-sales-expected-to-decline-7-2-year-over-year/.

4. Parks Associates, "TargetSpot Digital Audio Benchmark and Trend Study, 2012," http://www.targetspot.com/wp-content/uploads/2012/05/TargetSpot-Digital-Audio-Benchmark-and-Trend-Study_2012-White-Paper-copy.pdf.

5. Ben Sisario, "Radio Royalty Deal Offers Hope for Industrywide Pact," *New York Times*, June 10, 2012, http://www.nytimes.com/2012/06/11/business/media/radio-royalty-deal-offers-hope-for-industrywide-pact.html.

6. Trefis Team, "Sirius XM and Sound Exchange Square Off in the Battle over Royalty Rates," *Forbes*, April 10, 2012, http://www.forbes.com/sites/greatspeculations/2012/04/10/sirius-xm-and-soundexchange-square-off-in-the-battle-over-royalty-rates/.

7. Letter from Albert Foer and Diana Moss of the American Antitrust Institute to Congressman Dave Camp and Senator Max Baucus, posted at http://www.antitrustinstitute.org/~antitrust/sites/default/files/AAI%20Letter%20to%20Congress%20re%20Spectrum_0.pdf.

8. Bryce G. Hoffman, *American Icon: Alan Mulally and the Fight to Save Ford Motor Company* (New York: Crown Business, 2012), 1.

CHAPTER 9: AN ARMY OF NINJAS

1. Joel Levy, *Ninja: The Shadow Warrior* (New York: Sterling Publishing Company, 2008), 38.

2. David Jon Lu, *Japan: A Documentary History* (Armonk, NY: M.E. Sharpe, 1997), 203.

3. Philip N. Howard et al., "Opening Closed Regimes: What Was the Role of Social Media During the Arab Spring?," Project on Information Technology and Political Islam, September 2011, http://dl.dropbox.com/u/12947477/publications/2011_Howard-Duffy-Freelon-Hussain-Mari-Mazaid_pITPI.pdf.

4. Chris Taylor, "Twitter Users React to Massive Quake, Tsunami in Japan," *Mashable*, March 11, 2012, http://mashable.com/2011/03/11/japan-tsunami/.

CHAPTER 10: THE SHADOW WARRIOR

1. Stephen Turnbull, *Ninja: AD 1460–1650* (Oxford, UK: Osprey Publishing, 2003), 144.

2. Barton Whaley, in *Military History and Strategic Surprise*, eds. John Gooch and Amos Perlmutter (Totowa, NJ: Frank Cass, 1982).

3. Clayton M. Christensen, *The Innovator's Dilemma: The Revolutionary Book That Will Change the Way You Do Business* (New York: Harper-Business, 2011).

4. Brian Caulfield, "Steve Jobs Bio: Apple Has 3 Years of New Products in Jonathan Ive's Studio," *Forbes*, October 25, 2011, http://www.forbes.com/sites/briancaulfield/2011/10/25/steve-jobs-bio-apple-has-3-years-of-new-products-in-jonathan-ives-studio/.

5. Apple Inc., Form 10-K Annual Report SEC filing, October 26, 2011, http://biz.yahoo.com/e/111026/aapl10-k.html.

6. Matthew Shaer, "Newly Knighted, Apple Design Chief Hints at Mys-

tery Product," *Christian Science Monitor,* May 23, 2012, http://www
.csmonitor.com/Innovation/Horizons/2012/0523/Newly-knighted-
Apple-design-chief-hints-at-mystery-product.

7. Jason Chen and Jesus Diaz, "This Is Apple's Next iPhone," *Gizmodo,*
April 19, 2010, http://gizmodo.com/5520164/this-is-apples-next-iphone.

8. Barney Jopson, "Amazon Finds Upside to Sales Tax Payment," *Financial Times,* July 10, 2012, http://www.ft.com/cms/s/0/9bde8748-c201-11e1-8e7c-00144feabdc0.html#axzz21SISpQcy.

9. "Transmeta," Wikipedia, http://en.wikipedia.org/wiki/Transmeta.

10. Mark Henricks, "Loose Lips Sink Ships," CNN, June 19, 2011, http://
money.cnn.com/2001/06/19/sbstarting/entre_hush/index.htm.

11. Jenny Stover, "DataRPM Comes Out of Stealth Mode; Announces 'Instant Analytics' for Everyone," DataRPM press release, July 24, 2012,
http://news.yahoo.com/datarpm-comes-stealth-mode-announces-
instant-analytics-everyone-080243144.html.

12. Henry Clay Whitney, *Life on the Circuit with Lincoln* (Boston: Estes
and Lauriat, 1892), 178.

EPILOGUE: THE INTERNATIONAL CES KILLER STRATEGY

1. Joyce McKee, "Exhibitions Hold a Unique Value Proposition," *CEIR
Blog,* December 28, 2011, http://blog.ceir.org/ceir-reports/exhibitions-
hold-a-unique-value-proposition.

2. CEIR Index, An Analysis of the 2011 Exhibition Industry & Outlook,
available at: http://www.ceir.org/the_big_reports/exhibition_index/
exhibition_index.

3. See, for example, Gitomer's website, http://www.gitomer.com/.

INDEX